Essays on SIGMA References

Simulation and
Implementation of
General
Mathematical
Activity

John S N Elvey

Copyright © John S N Elvey

All rights reserved

ISBN-13: 978-1503051959

⟨## 1783 – 1902 : Math'l Economics⟩

δ: determinations/repr^{ns}/constructions   @: abstractions (? gen'z^{ns})
A: (non-)deterministic (max general) approximations
A: all other approaches (? ((q-)inverse) analogues

STUDIES in MATHEMATICAL ANALOGY/MATHEMATICS as
APPROXIMATION THEORY : Illustrative Material

A: $\delta \vee @ \vee A \vee A \vee \neg A$

δ 1. SAM routines for 'Schubert calculus'
   [Semple/Roth,Alg.Geom(OUP)/Kleiman(BAMS),for b/g.]
   [See also Proc AMS Symp PM XXVIII (2) 445-482 (SL Kleiman);
   LNM # 1462 (2) 235-252 (Ruud Pellikaan)]

δ 2. 'between-ness' and 'separation' axioms
   [JLMS 31(2)]

δ 3. DAC by matrix methods
   [JLMS 4/56240-248;ALSO: FEKETE,1916,H/L,etc.]
   ⟨Vermes, P., Amer. J.M. (1949) 541* Henrici Vol 1 §3.6

δ/A
4. Positive-def.fnctls on groups/approx.factorization
   [Hewitt/Ross,Abstract Harmonic Anal.,Springer,vol.2]
   (a) BONSALL/DUNCAN (Springer, 1973)*
   (b) C. BERG, et al. (Springer, 1984)* Z. SASVÁRI (AcadVerlag 1994*)

δ
5. Indiscernablility/exchangeability (L^p)
   [BAMS 83(1),121:ALDOUS,D.J.]

δ
6. Differentiation of algebraic fns.;extn to algebr.geom.
   [PLMS (3)9(1959)465-480:WALLACE,A.H.]

δ
7. Ergodic theorems for denumerable Markov chains
   [PCPS 63,pp 777-784]

δ 8. Use of non-C.A. set- fns for stochastic processes
   [PCPS 63,pp763-775]

δ 9. 'Factorizn at isolated points',for ODEs
   [PCPS 68,439-446]

δ 10. Between-ness and order in semigroups
    [PCPS 61,pp13-28]

A   11. Polynomials 'near to kth powers'
         [PCPS 61,pp1-5]

δ   12. A converse to Banach's FP theorem
         [J Res NBS B71(1967)73-   ,Meyers,P.R.]

A   13. Approx.of double limits by single limits/
         topology and convexity on {order classes}
         [J. IMA 3(1967)245-249]

δ/A 14. Distance functions for probability spaces
         [Renyi,A.,Th.Probability (N.-H.)p 74

δ/A 15. Proximity spaces as 'natural carriers of dynam.systems'
         [Hajec,O.,Dynamical Systems in the Plane(AP);idea only]

A/A 16. 'Conceptual extn'(VIA more general frameworks);cf,(D)AC
         [Examples to be furnished from any field ...]

δ/a 17. Extn of HOLDER inequality
         [General Ineqs 2. (=G2), (ed.,Beckenbach) 145-150] *

δ   18. Determination of a domain from its CENTRAL SET
         [G2,pp 357-366 (German) Milman]

δ   19. Derivation of all L(p) metrics from a 'probabilistic' one
         [G2,pp 429-434;Schweizer/Sklar]

A   20. Approximate n-connectedness
         [Genrl Topl.Appl.13(1982)225-228;CORAM,et al.]

δ 21. RECURRENCE in ergogic th.& combinatorial number th.
       [Princeton UP;FURSTENBERG,H.,et al.]

δ/A 22. { Number th.& algebra (many sharp results ...) }
       [Proc.,ZASSENHAUS,H.,ed.,AP,1977]

δ 23. Converse of Rouche's theorem
       [AMM (5/82,302-305);Estermann CV(Athlone P/London U)]

δ/A 24. Pointwise bounded density/distance estimates
       [TAMS ?p 62 GAMELIN/GARNETT] 143 (187-200) : MR 40 #...

δ 25. Constructive Riemann integration
       [PAMS (12/81,839-840)]

δ 26. Characteristic exponents/invariant manifolds
       [Ann.Math. 115(1982) RUELLE,D.] / Revs Mod Physics 57 (1985)
       617-656 (JP Eckmann/D. Ruelle)

A 27. Rates of convergence in 'weak laws of large numbers'
       [Ann.Prob.10(1982)374-381;HALL,P.ALSO:PITMAN Res.Notes]
       ALSO: LNM #861, 146-154 (V.K. Rohatgi) [2 refs.]

δ 28. Geometry of meromorphic functions   / GOLUZIN (Geom. F^n Th.
       [Mat.Sbornik 42(2) (1982) 155-197]   MR 36 #2973

δ/A 29. Bounds on various 'condition numbers'
       [Numer.Mat. 39,pp85-96] (Geurts, A.J.)

δ 30. 'Interface problems'   (H. HAN) / I(nfinite)EM (v. FEM)
       [Numer.Mat.39,pp39-50]

A 31. Approximation of the Hopf bifurcation (J. MARSDEN + (bk)×)
       Numer. Mat. 39 (1982) 15-37 (C. Bernadi)

[numer.Mat.39,pp 15-37]

δ/A 32. The 'gap between deterministic and stochastic DEs'
  [Ann.Prob.6,pp19-41(Sussmann)]

A 33. Best approximation of soln.to Ax=b,A>=0(matrix)
  [SIAM J Alg.Discrete Methods 3(1982)197-214]

δ 34. { The Stone-Cech compactification:characrn/embedding}
  [Ergebnisse Bd 83,Springer(WALKER,R.C.]

δ/A 35. Exchangeability (prob.th.)
  [Z.Wahrs.Th.Geb.60,pp249-281]

δ 36. Constructive results(functions of n variables)
  [Proc.Steklov Inst. 1981(4)]

δ/@ 37. Fixed-point theorems for product spaces
  [Pacific J.Math.99(2),(4/82)]

δ/A 38. Approximation theorems in shape theory
  [Topol.Applics.14(1982)111-116]

δ 39. Mappings ONTO metric spaces
  [Topl.Applics.14(1982)31-41]

δ/@ 40. Generalization of 'metric space'
  [Sci.Reps.Tokyo Kyoiku Daigaku 9(1967)236-254(Okuyama)]

δ 41. 'Representing systems'
  [Russian Math.Surveys 36(1) (1981)75-137,KOROBEINIK,Yu F]

δ 42. Existence of subspaces nonembeddable in sequence spaces
  [Dokl.24(2) (1981){:=D'81}202-206 (ANISKOVIC,E.M.)]

[D' ≡ Dokl.]

δ/A 43. Topol.properties of constructive plane curves /Paper by W. KAPLAN *
  [D'81 258-261,KUSNER,B.A.]

δ/A 44. Use of CHOQUET theory in mathematical economics
  [D'81 262-267,LEVIN,V.L.]

δ 45. Stability of CRAMER's theorem(math.stats.)
  [D'81 290-291,IL'IN,V.V.]

δ 46. Weakly suffic.sets/reprn of n-var.anal.fns by Dirichlet ser.
  [D'81,303-307,NAPOLKOV,V.V.,et al.]

A 47. Approx.of sum-distribs.by infinitely divis.laws(Levy metric)
  [D'81,382-386,ZAICEV,A.Yu.]

A 48. Approx.soln.of hydrodynamics eqs in n-connected domains
  [D'81,407-411,S.SMAGULOV]

δ/@ 49. Generalization of the'edge of the wedge theorem'('hyperfns')
  [D'81,425-430,ZARINOV,V.V.]

δ/@ 50. Ergodic theorems for nonlinear econometrics
  [D'81,430-433,ZAHAREVIC,M.I.]

δ/@ 51. Choice of metric for optimal machine models
  [D'81,434-437,MATUSOV,I.B.]

δ/A 52. Least-deviating exponential polns/opt.quadrature formulae
  [D'81,443-447,CAHKIEV,M.A.]

A/δ 53. Approximations for the measures of ergodic transns of R^1
[Z.Wahrs.Th.59(1982)27-38; has extn of Perron-Froben.thm]

A/δ 54. 'Laplace approximation' of sums of random variables
[Z.Wahrs.Th.59(1982)101-115]

δ 55. Computn of basis of syymetric fns in finite fields
[Math.Notes(Acad.Sci.USSR)30(1) (7/81)634-641,MIKHAILYUK]

δ 56. Embeddability of a locally normal group ...
[Math.Notes(Acad.Sci.USSR)(:=M)29(3)186-193,KURDACHENKO]

A/δ 57. Approx.of Lipschitz function by algebraic polynomials
[M 29(4)306-309,TEMHAKOV,V.N.]

A/A 58. Approximation of free groups w.r.t. conjugacy
[M 29(3) 196-198 (NEKRITSUKHIN)]

A/δ 59. Approxn.of functions in domains with quasi-conformal bdry
[M 29(3) 214-219 (ANDRIEVSKII,V.V.)]

δ 60. Series representation of entire fns rel.to $\{f(q_n z)\}$
[M 29(4) 258-265 (VINNITSKII,B.V.)]

δ/A 61. Differentiability of best-approximation operator/
Extremal interpolation/
Approximation of probability measures
[M 29 pp 295-320 KOLUSHOV,A.V.]

[M ≡ Math² Notes]

δ/A 62. Interrelations among'best approxns'for fns of n variables
[M 29(1) 51-58 (TEMLYAKOV,V.N.)]

A/@ 63. Rate of cgce in central limit th. for separable HS

[M 29(1) 79-82 (UL'YNANOV,V.V.)]

δ 64. Relations between 2-length,and,derived length (finite grps)
       [M 29(1) 85-90 (BRYUKHANOVA ,E.G.)]

δ/A 65. Power-bases for analytic functions
       [M 29(1)121-125 (OSKOLKOV,V.A.)]

A/δ/@ 66. Almost-sure summability of subsequences in BS
       [Studia Mat.71(1981)26-35/MPCPS 82,369-   (PARTINGTON)

δ 67. MS of relative order-type w(m)
       [Ann.Mat.Pura e Applic.128,pp 241-252]

A/δ 68. Entropy approach to stability of sampled data
       [Trans ASME J.Dynam.Systsems & Control 104(3/82)49-57]

δ/A 69. { Theory of Random Functions }
       [Pergamon,1965,pp 852;PUGACHEV,V.S.]

δ 70. { Compendium of continuous lattices (logic/algebra) }
       [North-Holland/?Springer (GIERZ,et al.+)]

A/δ 71. { Stochastic Analysis (bk)
       [KENDALL,D.,et al.,eds. (?AP,1969)]

A/@ 72. (i) Best approximation in von Neumann algebras
       (ii)General asymp.distrn for positive arithmetic fns
       [MPCPS 79(1)233-236;MPCPS 81(2) 43-54]

A/δ 73. A momotone convex analogue of linear algebra
   [Colloq.on convexity,Copenhagen,1965(:=C)261-276
       (Rockafellar)/AMS Memoir (Inst.)]

δ/A 74. Finite convexity in infinite-dimensional spaces
   [C 127-131 (HALKIN,H>)]

δ/@ 75. Finite convexity on NONconvex sets in inf.-diml spaces
   [C 20-33 (? HALKIN)]

A/δ 76. Analogues of Euler's 'polytope relations'
   [Convex Polytopes,by GRUNBAUM (Wiley) Ch 9]

δ/A 77. { Ergodic Theory and Information }
   [BILLINGSLEY,P.(Wiley,1965)]

δ/A/A 78. { Probabilities on Algebraic Structures }
   [GRENANDER,U. (Wiley,1963)]

A/A 79. { Stochastic Geometry }
   [KENDALL,D.,et al.,eds.,(?AP,1969)]

δ 80. { Riemann Surfaces }
   [GTM No.71 (Springer)]

δ/A 81. { Monotone Matrix Fns and Analytic Continuation }
   [DONOGHUE,W.F. (Springer,1974)]

δ/A 82. { Univalent fns and Conformal Mapping (extremal metrics) }
   { [JENKINS,J.A. (Springer)

δ 83. 'Finite sections of classical inequalities'
       [WILF,H.S. (Springer)] *

δ/A 84. Classifcn problems in ergodic theory(distinguishability)
       [PARRY/TUNCEL,LMS Lecture Notes No.67,CUP]

δ 85. { Classical and Modern Integration Theory }
       [PESIN,I.N. (Inst.)]

δ/A 86. { Probability Measures on Metric Spaces }
       [PATHASARATHY,K. (?Springer)]

δ/A 87. { Theories of Probability }
       [FINE,T.L. (AP,1973)]

δ 88. Axiomatization of 'multisets'(allowing repeats)
       [JLMS(3) 12 323-326]

δ/A 89. (i) Analogues of fields of Puiseux series
δ/A   (ii)Mathematical logic and Nevanlinna theory
       [(i) ?JLMS 8,504-506 ; (ii) IHES(lecture),1985+;
                                     Rubel,Lee A.]

A 90. Approximate Peano derivatives
       [JLMS 12,475-478]

δ/A 91. Condns for soln of $F(x,y,y',y'')=0$ to be $O(\exp\{Ax^m\})$
       [J.Math.Anal.Applics.78,497-521]

δ 92. Comparison of no.of solns of two systems of DEs
       [Diffl.Eqs(USSR) 16(1) (1980)1263-1270]

δ 93. Explicit soln of a first-order nonlinear(polyn.)ODE
       [Nonlin.Anal.5(2) (1981)157-165]

δ/@ 94. { Nonlinear AP Oscillations }*
        [KRASNOSELSKII,et al.,Halstead/Wiley,1973]

δ/@ 95. Stability theory for systems of infinitely many DEs
        [Tohoku Math.J.32,155-168]

A 96. Boundedness and asympt.vanishing of solns of 4th-order DE
       [Ann.Mat.Pura e Applic.123(1980)1-9]

δ 97. Condns for AP solution of (dx/dt)=f(t,x), (f periodic)
       [Bol.Un.Mat.Ital.A(5) 16(1979)412-417]

δ 98. 'Reconstruction theorems'in graph theory
       [Bull.LMS 14(3/4) (1982) ,NASH-WILLIAMS]

δ/A/A 99. (i) A bifurcation problem is PERSISTENT if g+ep='g
          (ii)A 'Tartar measure'with 'small support'is a POINT measure
          [See: Proc.NATO Instr.Conf.,Oxford,1983 ... (:= N)]

δ/A 100. CLAIM:A convex fn is finite at the bdry of a nonconvex set
         [See: N }

δ/A/A 101. Stability of the PROCESS of linearization
           [MARSDEN,J.E./CHILLINGWORTH,D. (REPORT,elasticity)]

δ/A 102. Liouville-type thm for systems of nonlinear PDEs
          (the only bounded ,global solutions are constant)
          [See: N (?GIUSTI/MEIER , ARANSON/CAFFARELLI)]

δ/A 103. Brownian motion and 'Picard's small theorem'
          [BAMS (7/82) 259-263]

δ/A/A 104. Noneuclidean functional analysis and electronics
          [BAMS (7/82) 1-64]

δ/A 105. (i) Inequalities for polynomials with a prescribed zero
δ/A   (ii) Constructive L(p)-approximation VIA analytic fns
          [Canadian J.Math.34,737-740; Ark.fur Mat.20(5/82)61-68]

δ/a/A 106. Local Paley-Wiener thm for noneuclidean Radon transforms
          [Comm.P&A Math.35,531-554]

δ   107. 'Calculus is algebra'
          [AMM 89(6),362-370;ALSO;KOK,A.,LMS Lec.Notes,CUP] *

δ   108. Completely and absolutely monotone functions
          [Canadian Math.Bull.25(2),143-148]

δ   109. Doubly-stochastic matrices as lin.combns of perm.matrices
          [Canadian Math.Bull.25(2),191-199]

δ/A 110. Inversion procedures for Radon transforms
          [AMM 89(6),377-384]

δ/A 111. 'Meaning and information in constructive mathematics'
       [AMM 89(6),385-388]

δ/A 112. Rearrangements of fns and bounds for eigenvalues of DEs
       [Applicable Anal.(:=AA)13(4),237-248]

δ   113. Hilbert transforms:extension of Gel'fand-Shilov techn.
       [AA 13(4),279-290]

δ   114. Nonuniqueness criteria for ODE x'=f(t,x),x(0)=q]
       [AA 13(4),291-296]

δ   115. The Poincare metric
       [Indiana U.Math.J.31(4),449-461]

δ   116. Extensions of triangular operators & matrix fns
       [Indiana U.Math.J.31(4),579-606 (DYM/GOHBERG)]

δ/A 117. (i)Polyn.bounds for orders of primitive solvable groups
       [J.Algebra 77,127-137]
       (ii)New criteria for solvability of finite groups
       [J.Algebra 77,234-246]

A/δ 118. 'Nearly-open'/'graph-closed' relations
       [Funda.Mat.114,219-228]

δ/A 119. Numerical procedure for conformal mapping
       [IMA J.Numer.Anal.2,169-181]

A/A 120. Comparison of fuzzy reasoning methods (??)
       [Fuzzy sets & systems 8,252-283]

A̸ 121. { Approximation of Real Functions }
      [Pergamon,1961 (TIMAN,A.F.;transl.)*]

δ   122. { Nonlinear Networks }
      [Elsevier,1978 (DOLEZAL,V.)]*

δ   123. Linear accessibility(for univalent functions)
      [JLMS (1973),385-398]

δ/A 124. Close-to-convex domains
      [Michigan Math.J.(1952),169-185]

?   125. UNIVERSALITY ('CHAOS',STAT.MECH.,...)
      [various collections of REPRINTS,etc..,...]

δ/A 126. Chordal metric
      [JLMS (1974),168-   ]

δ   127. Dimension-theory for s-frames
      [JLMS (1974)149-160]

δ/A 128. (i) Slowly-oscillating functions / JLMS(2)8(1974)99-108
      [Hardy,Divergent Series(OUP)]   [G E Peterson]
δ/A   (ii) Slowly-varying functions(KARAMATA)
      [Feller,Prob.Th.Vol2]

δ/A 129. Hilbert's inequality
      JLMS (1974),73-82]

δ 130. 'Pre-measures'
         [JLMS(1974),29-43]

δ/A 131. Fractional iteration
         [JLMS (1975),599-609]

δ 132. Distance and topology on cones
         [JLMS (1974),1-8]

A/A 133. Near-continuity
         [JLMS (1976),111-121]

δ/A 134. Amalgams of spaces
         [JLMS (1975),295-305]

δ 135. Integer-valued entire functions
         [JLMS (1975),501-512]

δ 136. Homeomorphisms of the plane with no fixed points
         [Abhan.Math.Sem.Uttamburg 30(1967)61-74]

A 137. Minimal spanning sets in complex approximation
         [JLMS (1976),317-322]

δ 138. Nonlinear ODEs On complex manifolds
         [JLMS (1976)258-262]

A/A 139. Approximate peano derivatives
[JLMS (1976)475-478]

A/A 140. 'Almost-complements' of subgroups
[JLMS (1963)761-768]

S/A 141. Inequalities for functions over collections of sets
[JLMS (1974)617-626]

S 142. Extension of Lipschitz (and other) functions
[JLMS (1974)604-608]

S 143. Restrictions on p-adic inntegrability
[JLMS (1974)731-734]

A 144. Approximation of convex/concave functions
[JLMS (1974)621-629]

S/A 145. Integration over vector lattices
[JLMS (1975)347-360]

S 146. Constructive extension of countable Boolean algebras
[JLMS (1975)337-346]

S/A 147. Ultimate boundedness of solns of 4th-order nonlinear ODEs
[JLMS 1975)313-324]

S 148. Conformal mapping of 'L-strips'
[JLMS (1975)301-307]

δ/A 149. Degrees of mapping:{homotopy classes}--> Z
        [JLMS (1974)385-392]

δ   150. Monodromy-preserving/spectrum-preserving deformations
        [Comm.Math.Phys.(1980)65-116]

δ   151. {Implementation of Schubert's enumerative calculus(SAM)}
        [Various sources:see,e.g.,1.,above]

δ   152. Explicit representation of $(d/dt)^n \{h(t)\}^k$
        [MPCPS (1978)]

δ/A 153. Invariant pseudo-metrics for co-set spaces
        [Hewitt & Ross,Abstract Harmonic Anal.Vol.1 para.8.14]

δ/A 154. { Prob.Th.:Independence,interchangeability,Martingales }
        [CHOW,Y.S.,et al.(Springer,1978),pp400]

δ/A 155. { Extremal Graph Theory }
        [BOLLOBAS,B.(AP,1978,pp488]

A/A 156. { Stochastic Dominance }
        [WHITEMORE,et al.,eds.(AP,1978),pp398]

δ   157. { The Anatomy of LISP }
        [ALLEN,J.R.(McGraw-Hill,1978),pp464]

A/A 158. Predominantly contractive mappings
        [JLMS 38(1963)81-86]

δ/A 159. Euclid's algorithm in algebraic number fields
     [JLMS (1963)55-59]

δ/A 160. Bounds for solutions of forced second-order ODEs
     [JLMS (1963)11-16]

δ 161. Equipotentials and Green's functions for convex domains
     [JLMS 30(1955)388-401]

δ/A 162. Mean-value theorems for infinite matrices in summability
     [JLMS (1958)50-62]

δ/A 163. Matrices with sets of prescribed eigenvalues
     [JLMS (1958)14-21]

δ/A 164. 'Hadamard multiplication' of entire functions
     [JLMS (1957)421-429]

δ/A 165. Directions of strongest growth of entire functions
     [JLMS (1957)409-420]

δ/A 166. Comparison of topologies
     [JLMS (1957)342-354]

δ/A 167. Theorems on compound convex bodies
     [PLMS (1955)358-391 ; JLMS (1957)311-318]

δ 168. Between-ness groups
     [JLMS (1957)277-285]

δ/ⓐ 169. Extension of Green's theorem (weak hypotheses)
[JLMS (1957)261-269]

δ/A 170. Existence of 'bounds' for infinite matrices
[JLMS (1957)203-213]

δ/ⓐ 171. Families of entire functions as entities (zeros)
[JLMS (1957)144-153]

δ/A 172. Bounds on diagonal elements of matrices with given E-V
[JLMS (1956)491-493]

δ/ⓐ 173. Riesz-almost-periodicity
[JLMS (1956)407-426]

δ/ⓐ 174. m-valued and ALEPH-0 -valued logics
[JLMS (1953)176-184]

δ 175. Function-theoretic methods for tauberian theorems
[JLMS (1953)94-102]

δ 176. Converse of Fatou's theorem
[JLMS (1953)80-89]

δ 177. Determination of an entire function by its Taylor series
[JLMS (1953)58-66]

δ/A 178. Inequalities for eigenvalues of matrices
[JLMS 28,8-20]

δ/ⓐ 179. Convolution of Cesaro-summability methods
[JLMS 30(1955)85--100]

δ/A 180. Boundedness of solutions of forced nonlinear ODEs
[JLMS (1955)64-67]

δ 181. Determination of general functions from Taylor series
[JLMS (1955)32-42]

δ/A 182. Proof of Cauchy's theorem VIA rectilin.approxns to curves
[JLMS (1955)1-11]

δ/ⓐ 183. Differentiability of norms of linear operators
[JLMS (1955)496-501]

A/ⓐ 184. Generalizations of arithmetic and geometric means
[JLMS (1955)449-463]

δ/ⓐ 185. Extension of Kakeya's theorem to zeros of infinite series
[JLMS (1955)314-319]

δ 186. Operator equations with Fredholm-type solutions
[JLMS (1954)318-326]

δ/ⓐ 187. Integration-by-parts for Cesaro-summable integrals
[JLMS (1954)276-292]

δ/A 188. Extremum problems for algebraic and trig.polynomials
[JLMS (1954)259-275]

δ 189. General classes of Prime-representing functions
[JLMS (1954)63-71]

δ/@ 190. Mean-value theorems in recursive function theory
[PLMS(2)52(2) 81-107]

A/@ 191. (i) Generalization of Minkowski's inequality
[PLMS 51(4)294-307]
(ii) 'Effectiveness at a point'of...polynomials
[PLMS 51,pp81-89]

δ/@ 192. Various species of INTEGRAL
[PLMS 52(5)365-377]

δ/@ 193. (i) Abstract Laplace transforms
[PLMS 54(2)94-111]
(ii)Galois group of binomial(explicit construction)
[PLMS (1953)195-211]

194. { Probabilities on Algebraic Structures }*
δ/A/@ [GRENANDER,U.,Wiley,1963]

δ 195. 'Partition calculus',in set-theory
[JLMS 40,pp137-148]

δ/@ 196. Generalized continued fractions
[KHOVANSKII,A.N.(North-Holland),Ch.4]

δ/A 197. A natural order-structure on C(E,F), (E,F,posets)
[JLMS 41(1966)323-332]

δ   198. Indecomposable positive,additive functionals
        [JLMS (1966)318-322]

δ   199. Generalization of the 'Riemann-Lebesgue THEOREM'
        [JLMS (1966)313-317]

δ/A 200. Diophantine equations/elliptic curves
        [JLMS (1966)193-292 (CASSELS)]

δ/A 201. Limit-distance of Hausdorff transns of tauberian series
        [JLMS 40(1965)295-302]

δ/A 202. Theorems on 'contractive maps'
        [JLMS (1966)101-106]

@   203. The role of algebraic topology in mathematics
        [JLMS (1966)63-69 (ATIYAH)]

δ   204. Methods for entire and meromorphic functions
        [JLMS (1966)11-28 (NEVANLINNA)]

δ   205. { Formes Exterieures et leures Applications }
        [SLEBODZINSKI,W.(PWN,Warwsaw,1963)]

δ   206. Element-distibution in infinite bi-stochastic matrices
        [JLMS (1965)689-698]

δ/A 207. 'Continuous analytic continuation'
        [Nonlin.Diffl.and Integral eqs.(DAVIS,H.T./Dover)*]

δ/A 208. Convexity in (Wald's) statistical metric spaces
[JLMS (1963)117-128]

δ/A 209. Algebraic approximation of manifolds
[PLMS (1957)196-211 (WALLACE,A.H.)]

δ/A 210. Normal-mode vibrations for nonlinear systems
[PCPS 60(1964)595-611]

δ/A 211. Central limit theorem for proc.on finite Markov chains
[PCPS (1964)547-567]

A/δ 212. Finite approximations to infinite nonnegative matrices
[PCPS (1968)465-470 ...]

δ 213. Smale's formulation of electrical network theory
[MPCPS 79(1979)537-544] / QAM (1964): BRAYTON/MOSER *
I, II

δ/@ 214. Axiomatics of 'nearness' and 'compression'
[MPCPS (1979)469-471]

A/@ 215. Central limit theorem for BS-valued RVs
[Advances in Probability(No.4);Dekker]

δ/@ 216. 'Dentability' for BS-valued martingales
[LNM 472 (WOYIZINSKI)]

A/@ 217. Banach distance $d(E,F) = \inf_{I,I^{-1}} \{\|I\| \|I^{-1}\|; \{I\} \equiv \{iso, E \to F\}\}$
[LNM #672, p278;ALSO:books on functional analysis]

A 218. Approximation of 'countable'by 'finite,'for Markov chains
[FREEDMAN,D. (Holden Day,1972) }

A/@ 219. (Uniform) Stochastic continuity on a compact set
[Gihman-Skorohod (Springer),Vol 1,p169]

S/A 220. Quasi-inverse in a Banach algebra without unit element
[Hille-Phillips :Fnl Anal.and Semigroups (AMS,1957)]

S/@ 221. Generalized inverses (of linear operators ...)
[e.g.,NASHED,M.Z.(AP)]

S/A 222. Martingales in sigma-finite measure spaces(analogues)
[TAMS 97(1960)254-285]

A/A 223. One-sided analogue of the Kolmogorov inequality (stats)
[Ann.Math.Statist.31(1960)483-497]

A/@ 224. Multivariate Chebychev inequalities
[Ann.Math.Statist.(1960)1001-1014]

S/@ 225. Martingales of BS-valued RVs
[BAMS 66(1960)395-398]

S 226. Homology theory VIA pairs of cells with 'facing relations'
[PLMS 12(1962)609-690]

S 227. Quadratic-form transns that do not increase the class nos.
      [PLMS (1962)577-587]

     228. Summability of Laplace-Stieltjes integrals
S/@   [PLMS (1953)267-304 ; PLMS (1961)654-690]

     229. (i) Distribution of fixed points of entire functions
A/S   [PLMS (1966)493-506]
        (ii) 'Generalized functions' for (l.)c.(Abelian) groups
S/@   [PLMS (1966)415-456]

A/@ 230. (i) Probabilistic generalization of 'large seive'
         [RENYI,A.,Prob.Th.(N.-Holland),p285]
 A   (ii) Quasi-orthogonal functions
         [RENYI,A.,Prob.Th. ...]

S/A 231. Martingale formulation of ergodic theorems
         [PAMS 10(1959)531-539]

S   232. The space of subspaces of a BS
         [JLMS (1972)67-73]

S/A 233. Various (equivalent) charactrzns. of quasi-uniformities
         [JLMS (1972)48-52]

S/A 234. Extensions of continuous functions to compactifications
         [JLMS (1972)15-20]

S   235. Deterministic nonstationary sequences of RVs
         [JLMS (1978)187-192 ;PLMS (1973)485-496]

A 236. Approximate identities in normed algebras
[JLMS (1978)141-151]

S 237. Deduction of 'function-class of B' from 2nd-order informn.
[JLMS (1978)102-106]

S 238. Reconstruction of families of sets ...
[JLMS (1978)5-9]

S/A 239. Convexity in statistical MS
[JLMS (1964)117-128]

S/A 240. Metrics for Wald's(statistical metric) spaces
[JLMS (1964)129-130]

S/A 241. Measure of asymmetry for plane convex sets
[JLMS (1964)95-102]

S 242. Interrelations of 'Abel'and 'Riemann' summability
[JLMS (1964)5-11]

S/A 243. Categories of information (mechanics)
[Contributions to Mechanics (ed.,ABIR)pp87-100]

S 244. Integrity bases for N symmetric matrices
[Contrib.to Mech.(ed.,ABIR)121-141]

S 245.(i) Flow-patterns around small/thin bodies ...
S [Contrib.to Mech.(ed.,ABIR)205-242]
S (ii)Rheology of complex bodies

[Contrib.to Mech.(ed.,ABIR)391-436]

δ/A 246. Manipulation of chemical formulae in computers
     [Comptr.Techn.Chem.Res.(bk); CHEMTRAN (MIT)]

A   247. (i) Majorization:inequalities(prob./stats.)
         [MARSHALL/OLKIN (AP,1979)pp569]
         (ii) Inequalities foir multivariate distributions
         [TONG (AP,1980);BAMS (11/81)319-324]

δ/A 248. Complexity aspects of the 'fundamental theorem of algebra'
         [BAMS (1/81)1-36 (SMALE)]/Constr. Aspects of FTA(bk)*

δ/A 249. Boundaries induced by nonnegative matrices
         [TAMS 83(1956)19-54]

δ   250. Spaces associated with Markov chains
         [PLMS (1960)67-87 (KENDALL,D.G.)]

δ/A 251. Positive-definite functions of a real variable
         [PLMS (1960)53-66]

δ/A 252. Summability of Cauchy products of series
         [PLMS (1959)556-574]

δ/A 253. Cesaro-Perron(C.-P.) AP functions(/distance)
         [PLMS (1952)150-174]

A   254. Boundedness criteria for various sequence spaces
         [PLMS (1965)422-436]

δ/@ 255. 'Stationarity' for abstract-valued procedures
[PLMS 15,pp385-398]

δ/A 256. Fractional differences and summability
[PLMS 1963,pp430-460]

A/@ 257. Generalizations of 'Chebychev inequalities'(moments)
[PLMS 13,pp385-412]

δ 258. LIMIT-CIRCLE property holds for ALL q,if for any q ...
[CODDINGTON/LEVINSON,Th.of ODEs,p225]

δ/@/A 259. Special forms of divergence/singularity for TS
[PLMS 12,pp121-132]

δ 260. Representability of lattice groups by real cts fns
[PLMS 12,pp100-120]

A/δ 261. Effective sets/range of effectiveness(for germ spaces)
[PLMS (1968)745-767]

A 262. Asymptotic properties of linear operators
[PLMS (1968)405-427]

δ/A 263. Convexity properties of strong Riesz-summability (DS)
[PLMS 18,pp367-384]

A 264. Real zeros of random polynomials
[PLMS 18,pp29-35 ; PLMS 18,pp308-314]

265. Extension of theorems for monotone oprs to general oprs
S/@ [PLMS 20,pp451-468]

266. Multidimensional forms of the Berry-Esseen theorem(prob)
S/A [PLMS 20,pp33-59]

267. Equivalent circuits in electromagnetism/electronics
S/A [e.g., Mech.Elec.Vibrations (BARKER,J.R.)*,Methuen]

268. Abstract Riemann surfaces of fields
S/A [Zariski-Samuel,Commutative Alg.,Vol 2;Saks-Zygmund]

269. Valuation-theoretic(metric)concepts for fields
S/A [Zariski-Samuel,Comm.Alg.Vol 2]

270. Solution of infinite linear alg.systems by truncation
A [PCPS (1968)1215-1221]

271. Stochastic analogues in quantum mechanics
A [PCPS 1968,pp1061-1070]

272. Generalizations of the Holder and Minkowski ineqs.
A/@ [PCPS 1968,pp1023-1027]

273. Generalizations/applications of Cauchy's inequality
A/@ [QJM(Ox) 40(1967)247-250 ; J.Math.Anal.App.12,pp491-494]

274. Extension of results on T(V)S to general TS
δ/𝔄   [PCPS (1968)965-973]

275. Approximate factorization of residual likelihood criteria
𝔄   [PCPS 62,pp743-752]

276. Interpolation theorems on entire functions('Liouville')
δ   [PCPS 62,pp721-742]

277. 'Supermultiplicativity' of nonnegative permanents
δ   [PCPS 62,pp643-648]

278. Complete subsets of mappings over finite domains(logic)
δ   [PCPS 62,pp597-611]

279. 'Structure of X' VIA study of {subsets of X}
δ   [PCPS 62,pp583-595]

280. NONrepresentability of HEUN fcns as contour- integral solns
δ   [PCPS 65,pp447-459]

281. Bounds for solutions of the hyper-elliptic equation
δ/𝔄   [PCPS 65,pp439-444]

282. Dissection of a rectangle into squares or rectangles
δ   [PCPS 65,pp399-408]

283. 'Ignorable co-ordinates'(dependence)
δ   [e.g.,WHITTAKER,E.T.,Analytical Dynamics(CUP)]

284. Effect of boundaries on finite Markov chains
δ/A   [e.g.,Feller,Vol 1; PCPS 61,pp173-190]

285. 'Games on TS'
δ/A   [PCPS 61,pp165-171]

286. Explicit solutions of undamped Duffing ODE
δ   [e.g.,PCPS:51,pp297-312;54,pp426-438;61,pp133-164]

287. Continuity of lacunary structure
δ   [PCPS 61,pp109-112]

288. Approximation in measure, of G(d)-sets by F(s)-sets
δ/A   [PCPS 61,pp105-108]

289. Between-ness in semigroups
δ/@   [PCPS 61,pp13-28 ; JLMS (1957)277-285]

290. Repesentation of function varieties by expontl polns
δ   [PCPS 61,pp395-424]

291. Differentiability/analyticity on topol.groups
δ/@   [PCPS 61,pp347-379]

292. Waring's problem in algebraic number fields
δ/@   [PCPS (1960)449-459]

293. n-variable forms of Lagrange's(reversion)expansion
δ/@ [PCPS:56,pp366-380;61,pp499-517]

294. Aggregation (clustering)
A [PCPS (1965)475-498]

295. Hermite-polynomial expansions of generalized functions
δ/A/@ [PCPS 68,pp129-139]

296. Rational fcns with spec.zeros/poles over algebraic curves
δ/@ [PCPS 68,pp105-123]

297. Various definitions of order-continuity between TS
δ [PCPS 68,pp27-31]

298. 'Topologies of separate continuity' on TS
δ [PCPS 68,pp663-671]

299. Lefschetz fixed-point theorems for multivalued maps
δ/@ [PCPS 68,pp619-630]

300. Specialization/conservtn- of- no.of solns({eqs in rings})
δ [PCPS 69,pp59-70]

301. Ternary algebras(operations on sets of triples)
δ [PCPS 69,pp25-52]

302. On hearing the shape of a drum
δ [PCPS 69,pp353-363 (?KAC)]

303. Specification of filters by response functions
S/A [PCPS 63,pp221-227]

304. Inequalities involving Radon-Nikodym derivatives
A [PCPS 63,pp195-198]

305. Generalizations of the Laplace transform
S/@ [PCPS 63,pp155-160]

306. Countable mixtures of discrete probability distributions
S [PCPS 62,pp485-494]

307. Consistency/relative strength for summability methods
S/A [PCPS 62,pp421-428]

308. Schauder decompositions of l.c.TS
S [PCPS 68,pp377-392]

309. General theorem implying both Tikhonov & Ascoli thms
S/@ [PCPS 68,pp351-354]

310. Non-repeating(e.g.,Morse-Hedlund)sequences
S [PCPS 68,pp267-274]

311. Gaps and steps in {nq(mod 1)}
S [PCPS 63,pp1115-1123;ALSO:KUIPERS/NIEDERETTER,bk]

312. General theory of topological sequence spaces
    δ   [PCPS 63:pp963-981 & pp997-1019]

313. Finite approximation to infinite nonnegative matrices
    A   [PCPS 63,pp983-992]

314. Waring's problem for homogeneous forms
    δ   [PCPS 65,pp663-672]

315. Riemann surfaces as TSs associated with field extensions
    δ/A [PCPS 65,pp635-650]

316. Convergence-regions of simple Taylor/Dirichlet series
    δ   [PCPS 65,pp619-634]

317. Bounds for characteristic exponents of $x' = A(t)x$
    A   [PCPS 61,pp889-896]

318. Generalization of 'arc-connectedness'
    δ/a [PCPS 61,pp879-880]

319. Connection, invariance, and stability (for flows)
    δ   [PCPS 60,pp52-55]

320. Relative simplicial approximation
    A   [PCPS 60,pp39-43]

321. Ergodic theorems for {infinite stochastic matrices}
    δ/A [PCPS 63,pp777-784]

322. Almost-subharmonic functions
 A/@ [RADO,T.,Subharmonic functions (bk)p26-]

323. Multivariate probabilistic inequalities
 A [BAMS (11/81),review of bk(Tong,et al.,AP,1980)*]

324. 'Stochastic intervals'
 A/@ [MEYER,Prob.and Potentials (Blaisdell)]

325. Sequential convergence in l.c.TSs
 δ/@ [PCPS 64,pp341-364]

326. Proof of Routh stability criterion VIA Liapunov stability
 δ [PCPS 58,pp694-702]

327. Some 'pathological plane curves'
 δ [PCPS 58,pp569-574]

328. (i) Eigenvalues of composite matrices
 δ [PCPS 57,pp37-49]
    (ii) Metrical properties of compound matrices
 δ/A [J.Math.Anal.App.(?1961),OSTROWSKI,A.M.]

329. Distribution of curvature on random Gaussian surfaces
 δ/A [MPCPS 54,pp439-453]

330. Stabilization of matrices
 δ [MPCPS 54,pp417-425]

331. 'Schubertian computations'
   δ   [MPCPS 54,pp399-416]

332. Inequalities of Chebychev type
   A   [MPCPS 59,pp135-146]

333. Preference-order inequalities
   A   [MPCPS 59,pp125-133]

334. Exact solution of an ODE VIA approx.soln of other ODEs
   δ/A   [(M)PCPS: 49,601-611;57,790-810;59,95-110]

335. Approximation of integrals of periodic/rapidly-oscl fns
   A   [MPCPS 59,81-88]

336. Entire-function summability methods
   δ/A   [MPCPS: 55,23-30;56,125-131]

337. Generalization of Wald's identity for Markov chains
   δ/@   [MPCPS 56,205-214]

338. Generalizations of Gateaux,and Frechet, derivatives
   δ/@   [?Federer,Geometric Measure Th.(Springer)]

339. Radiation from a random surface
   δ/A   [MPCPS 56,27-40]

340. 'Random flights' for small numbers of vectors
   δ/A   [MPCPS 56,21-26]

341. Accessibility and metrical transitivity(topol.dynamics)
   δ   [MPCPS 57,833-842]

342. Use of generalized functions to solve integral equations
   δ/@   [MPCPS 57,767-777]

343. Rate of approximation by 'polyns'with complex exponents
   A   [JLMS(2)15,305-318]

344. Polynomial approximation in Bers spaces
   A   [JLMS 15,255-266]

345. 'Effectiveness'of basic sets of holomorphic functions
   δ/A   [JLMS 15,267-270]

346. Self-conformal maps of the Unit Disc
   δ   [JLMS 15,239-254]

347. Hausdorff distance
   δ/A   [e.g.,DUNFORD/SCHWARTZ]

348. Examples of NON-pseudometrizable TSs
   δ   [General Topol.(CSASZAR,A.),Hilger,1980,pp87-]

349. Neighbourhood structures/spaces
   δ   [e.g.,CSASZAR,General Topol.,pp60-62]

350. A (non-metric) size fcn: $D(A,B):=\inf\{d(x,y):x \text{ in } A, y \text{ in } B\}$
   δ   [e.g.,CSASZAR,GT,for discussion of similar examples]

351. Nerve of a covering of a TS
δ/A [e.g.,NAGATA,Modern GT,p301-]

352. 'Compact consequence relations'
δ/@ [Multiple-Conclusion Logic'(SHOESMITH/SMILEY),CUP,1978]

353. Detailed calculations in homotopy theory
δ [e.g.,Elements of HT (WHITEHEAD,G.W.),Springer,1978]

354. Topological degree (?as a generalization of 'residue')
δ/@ [e.g.,Degree Theory (LLOYD,N.),CUP]

355. Approxn of sums(indt RV)by infinitely-divisible distbns
A/δ [Sankhya A26(1963)159-174]

356. 'New methods of approximation' in theory of Markov chains
A/δ [PLMS 16(1966)213-240]

357. 'Stochastic geometry'
@/A [Stochastic Geometry (KENDALL,D.G.,ed.),Wiley,1974]

358. Generalization of 'fundamental theorem on subadditive fns'
δ/@ [PCPS 58(1962)235-238]

359. Hausdorff measure of {planar Brownian paths}
δ/A [PCPS 57,209-222]

360. Brownian-motion proofs of classical analytical results
δ/A/@ [KAHANE,J.-P.],?BAMS (/SW)]

361. Characterization of stochastic proc.VIA covariances
   S [e.g.,LAMPERTI,Stochastic Proc.,Springer,for examples]

362. Orthogonal increments (of random processes)
   S [e.g., LAMPERTI,SP,p22-]

363. Deterministic/stochastic approximation of regions
   A/S [DAVIS,P.J.,J.Approx.Th. (/SW)]

364. Dimension of a proximity space (PS)
   S [AMS transl.(2)21(1962)1-20]

365. Completions of PSs
   S [AMS transl.(2)38(1964)5-73 (SMIRNOV,Yu.M.)]

366. Generalized functions in infinite-dimensional spaces
   S/@ [Trans.Moscow Math.Soc.,1971-1974+]

367. BV-problems for the Laplacian in infinite-dimensl spaces
   S/@ [Trans.Moscow Math.Soc.(various)]

368. 'Clutters'(collections of'incomparable sets')
   S/@ [JLMS 12,356-360]

369. Topological groups assoc.with species of interval analysis
   S/A [RAAG Mem.(Inst.)]

370. TS and statistical MS
   S/A [Gen.Topol.Appl.9(1978)233-237]

371. { Survey of Dimension Theory --to 1966 }
   δ   [Prague Symposium ... (/SW)]

372. { Survey of Dimension Theory --to 1970 }
   δ   [Gen.Topol.Appl.1(1971)65-77]

373. Orderable TS
   δ/A [Gen.Topol.Appl.2,1-10]

374. A concept of nearness
   δ/A [Gen.Topl.Appl.5,191-212 (HERRLICH,H.)]

375. Statistical MS
   δ/A [Pacific J.Math.(1960)313-334 (SCHWEIZER/SKLAR)]

376. Topological properties/B-completeness in NSs
   δ/A [Gen.Topol.Appl.:5,263-268;8,111-118]

377. Descriptive complexity of functions
   δ   [KATETOV,Prague Symposium,214-219 (/SW)]

378. Distance functions into (partially-)ordered sets
   δ/A [REICHEL,H.-C.,papers,and Annotated Biblio.(Vienna)]

379. 'Generalized sequences' for finite-difference calculus
   δ/@ [J.SIAM ?1965)177-200 (TRAUB,J.F.)]

380. Binding spaces(unifying completion and extension)
[Funda.Mat.76(1972)43-61*]

381. Proximity spaces (general treatment)
[NAIMPALLY/WARRACK;CUP]

382. Contiguity spaces
[Izv.Akad.Nauk SSSR 23(1959)613-634 (IVANOV(A))]

383. Contiguity of probability measures
[ROUSSAS;CUP]

384. Homomorphic approximation to BV algebras
[BIRTEL,F.T.;BAMS 84,406-]

385. Combinatorial inequalities and smoothness of functions
[GARCIA,A.M.;BAMS 82,157-170]

386. Intrinsic distances
[KOBAYASHI;BAMS 82;357-416]

387. Approximation-solvability of operator equations
[BAMS 81,223-312;PETRYSHYN]

388. Reconstruction of objects from radiographs
[BAMS:83,1227- ,84,691-]

389. Topological definition of 'smallness of a set'
[BAMS 72,579-600;DOOB,J.L.]

390. Statistical geometry
     [BAMS 79,829-856;GRENANDER,U.]

391. Norms and localization of roots of matrices
     [BAMS 74,816-830;HOUSEHOLDER,A.S.]

392. Best approximation in $L^1(T)$
     [BAMS 80,788-804]

393. Mechanized mathematics
     [BAMS 72,739-750;LEHMER,D.H.]

394. Metric entropy and approximation
     [BAMS 72,903-937;LORENTZ,G.G.]

395. Applications of quasi-isometric measures
     [BAMS 76,427-528;MASANI,P.]

396. Developments in Schauder-basis theory
     [BAMS 78,877-908;McARTHUR,C.W.]

397. Almost-normal operators
     [BAMS 79,615-624;PUTNAM,C.R.]

398. Survey of integral-representation theory
     [BAMS 76,159-227;REINER,I.]

399. Quasi-subordination/majorization(coeff.conjectures)
     [BAMS 76,1-9;ROBERSON,M.S.]

400. Method of extremal length
   S/A   [BAMS 80,587-606;RODIN,B.]

401. Algebraic integration theory
   S   [BAMS 71,419- ;SEGAL,I.]

402. Characterizations of linear groups
   S   [BAMS 75,1043-1091;SUZUKI,M.]

403. Functions of several noncommuting variables
   S/@   [BAMS 79,1-34;TAYLOR,J.L.]

404. NONsolvable finite group with all local subgroups solvable
   S   [BAMS 74,383-437;THOMPSON,J.G.]

405. (Inter)connection problems for asymptotic series
   S/A   [BAMS 74,831-853;WASOW,W.]

406. Open questions on singularities
   S   [BAMS 77,481- ;ZARISKI]

407. Sobolev spaces ... and perfect splines
   S/A   [BAMS 79,124-128 (KARLIN,S.)]

408. Inequalities for uniformly bounded dependent variables
   S/A   [BAMS 79,40-44 (FREEDMAN,D.)]

409. A new form of compactness for TSs
   S   [Funda.Mat.66(1969/70)185-193]

410. Generalized convex functions
δ/@ [TAMS 58(1945)220-230]

411. 'Principle of nonsufficient reason' (for inequalities)
δ/A [Inequalities (?Vol.1) (:=I)SHISHA,O.,ed.(AP,1967)1-15]

412. Extension of Minkowski's inequality
δ/A [I,p39,Sec.2]

413. Kantorovich inequalities
A   [I,p45,Sec.3]

414. 'Musical scores of time-functions'&'uncertainty relations'
δ/A [I,pp57-71]

415. Variants of Wirtinger's inequality
A   [I,pp79-103]

416. Applications of 'majorization'
A   [I,pp177-203;ALSO: MARSHALL/OLKIN (bk)]

417. Elimination for algebraic inequalities
δ/A [BAMS 61,326 (MOTZKIN,T.S.);bk:Springer ...]

418. Approximation of domains of holomorphy
δ/A [Lecs.Fns CV, KAPLAN,W.,ed.(U.Michigan P,1955)pp358-]

419. Metrization of domains of holomorphy
δ/A [Lecs.fns CV,KAPLAN,W.,ed.,pp375-]

420. Definitions of boundedness in general TSs
δ/A [Gel'fand/Shilov,GFs,Vol.2,p31,ff;HU,S.-Z.(paper/SW)]

421. The symmetric group as a MS
δ/@ [JLMS 35(1960)215-220]

422. NON-continuable power-series
δ   [JLMS 35(1959)117-127]

423. SUB-stochastic matrices
δ/A [JLMS 41(1966)605-611]

424. Contractive mappings (Of various types)
δ   [JLMS 41(1966)101-106]

425.(i) Entire and meromorphic function(various techniques)
δ   [JLMS 41(1966)11-28]
    (ii){ Recent Advances in Matrix Theory }
δ/A [SCHNEIDER,H.,ed.,U.Wisconsin P.,1964,pp142]

426. Extensions of the Fredholm scheme for integral equations
δ/@ [PLMS (3)1(1951)327-384]

427. 'Steiner symmetrization' in proofs of inequalities
δ/A [JLMS 32(1957)102-108;ALSO POLYA,in:Maths.for Engrs 2 ...]

428. Hilbert inequalities for det{1/(m+n+q)}
A   [JLMS 32,7-17]

429. Estimates of the rank of circulants
    δ/A [JLMS 31(1956)445-460]

430. Various classes of AP functions
    δ   [JLMS 31,407-426]

431. Between-ness and separation
    δ   [JLMS 31,240-248]

432. Explicit calculation of 'extemal polynomials'
    δ/A [JLMS 31,191-199]

433. Spectra of infinite matrices VIA finite sections
    δ/A [JLMS 32,357-367]

434. Comparison of topologies
    δ   [JLMS 32,342-354]

435. 'Compounds' of convex bodies
    δ   [JLMS 32,311-318]

436. Between-ness groups
    δ   [JLMS 32,277-285]

437. BSs of distributions(defined VIA strong convergence)
    δ/@ [JLMS 32,483-498]

438. Sharp bounds on mod(f'(z)),for bounded,regular f
    A   [JLMS 32,430-435]

439. Hadamard multiplication for entire functions
   S   [JLMS 32,421-429]

440. 'Isolability'of singular points in DAC
   S   [JLMS 32,409-420]

441. Properties of A^{-1} in relation to 'summability by A'
   S/A   [JLMS 32,397-408]

442. Finite-dimensional subspaces as 'negligible sets'
   S/A/A   [e.g.,results in spectral theory;DIEUDONNE(FMA)]

443. An 'implicit fcn theory' without differentiability conds.
   S   [Math.Zeit.89(1965)137-180] QUADE, et al.. MR 31 #3557

444. Choquet boundary theory for posets
   S   [Studia Math.36(1970)177-193]

445. Liouvill's(CV)theorem as a result on topol.of vector fields
   S/A   [e.g.,MAA Studies,#4(Global Anal),CHERN,S. S.,ed. ...]
   and (for DFS; Morse, M. 'Topol Props. in Fns CV (bk))

446. Guidance systems and 'apparent interaction'
   S/A   [KORENEV,Mechs.of Guided Bodies,p70]

447. General notions of 'constraint'
   S   [KORENEV,Mechs.of Guided Bodies,Sec.1.2]

448. Characterization problems in mathematical statistics
   S/A   [KAGAN/LINNIK/RAO,Wiley,1973;pp499]

449. Characterizations of the normal distribution
S   [MATHAI/PEDERZOLI,Wiley,1977,pp149]

450. Normal(Gaussian)approximations and asymptotic expansions
S/A [BHATTACHARYA/RAO,Wiley,1976;pp274]

451. Iconic calculus -- symbols with feeling !!
A-∞ [General systems (RAPOPORT,A.,ed.)20,71-93]

452. Some general operations for matrices(e.g.,differentiation)
S   [General Systems 20,145-150]

453. Embeddability and the 'word problem'
S   [JLMS 28,76-80]

454. Recursive continuity
S/A [pp321-329 in Construc.Aspects F Thm Alg(DEJON,et al.)]

455. { Computable Analysis }
S   [ABERTH,O.,McGraw-Hill,1980]

456. Inequalities related to p-groups
A   [PLMS (1960)24-30]

457. Positive-definite functions on R
S   [PLMS (1960)53-66]

458. Spaces associated with Markov chains
   δ   [PLMS (1960)67-87]

459. Harmonic functions and tauberian theorems
   δ/A [PLMS (1960)88-106]

460. Convergence factors for 'Ward integration'
   δ   [PLMS (1960)107-121]

461. H^p and L^p as linear TSs
   δ/@ [GAMKRELIDZE,R.V.,ed.,Progress in Math.1,pp114-]

462. Extremal polynomials associated with plane-curve systems
   δ/A [Advances in Math.3,137-232 (WIDOM)]

463. Abstract algebraic structures in beam design
   δ/A [Acta Tech Sci Hungar ?91(1980)355-371]

464. Computation of the Zeta function and related functions
   δ   [USSR Compl Math Math Phys 20,pp212-]

465. Characterization of {limit- points of zeros of annular fns}
   δ   [J fd R u A M 246(1971)202-]

466. Analytic functions in TVSs
   δ/@ [Studia Math.37(1971)77-]

467. Nonnegative matrices
   δ/A [Pacific J.Math.36(1971)755-]

468. Analytic functionals
   δ/a [J.Sci.Hiroshima 34(1970)145-]

469. Functional eqs. for linear combins.of powers of functions
   δ [Tohoku Math.J.23(1971)289-]

470. Analytic functionals
   δ/a [Michigan Math.J.18,141;J.Math Anal Appl 34(1971)147-]

471. Zeros of analytic functions
   δ [J Math Anal Appl 35(1971),YANG,C.-C.]

472. Power-sum product expansions
   δ [Duke Math J 38,pp285- (RIORDAN,J.)]

473. Solvable 'cell model'
   δ [J.Math Phys 12(1971)766-]

474. Chebychev approximations for the Zeta function
   A [Math.Computn.25(1971)537-]

475. Average order of arithmetic functions
   δ/A [J.Number Th.3(?1971)184-]

476. Stability/bifurcation
   δ/A [IOOS/JOSEPH,Springer,1971]

477. { Value-Distribution Theory }
   δ/A [STOLL,et al,Parts A,B;Dekker,1974]

478. Nonlinear operator theory
   S/A/@ [ZEIDLER,Vols 1-5 (Springer);HUTSON/PYM (intro.)]

479. Multi-component random processes
   S [Advances in Probability,Vol 6 (Dekker,1980)]

480. Metrics in BSs
   S/A [FRAZONI,et al.,eds.N>-Holland Math Stud.40:Holo.Maps...] ✷

481. { Proc.1981 MACSYMA Users'Conference } ✷
   S/A [WANG,P.S.,ed. (ACM,NY)]

482. Algorithms for first-/second- order ODEs
   S/A [PRELLE,M.J.,PhD Thesis(RPI);KOVACIC,J.J.(preprint)]

483. { Topology of Uniform Structures }
   S [PAGE,W.,Wiley;1978 (SEE BAMS for review)]

484. ? Symbolic computation in optics
   S/A [BUCHDALL,books;Dover,1978,CUP,1970]

485. Analytic-function methods in probability theory
   S [GYIRES,ed.,Colloq.J.Bolyai,No.20]

486. Mathematical theory of entropy
   S/A [MARTIN/ENGLAND,Encycl.Math.,Vol 12;A-W/CUP]

487. { The Algebra of RVs (integral-transform methods) }
δ   [SPRINGER,M.D.(book)]

488. { Computers and Intractability }
δ/A [GARY/JOHNSON (book)✕]

489. { Contemporary Complex Analysis }
δ   [BRANNAN/CLUNIE,eds.,AP,1980]

490. { Topics in Topology }
δ/A [CSASZAR,A.,ed.;Colloq.J.Bolyai No.23 }

491. { Bounded Analytic Functions }
δ/A [GARNETT,J.;AP,1980]

492. Homotopy methods for fixed points
δ/A [GARCIA/ZANGWILL,P-H,1980]

493. Automatic differentiation (for optimization;not SAM)
δ/A [Springer LNCS No.120]

494. Boolean techniques in urban design
δ/A [RENE/THOMAS,eds.,LNP(hysics)]

495. CT-modelling in urban design
δ/A [WILSON,?R.;U.Calif.P.,1980]

496. AUTOMATH (proof-checking)
δ   [REZUS,Abstract Automath(Math.Centrum Tract/miscl.repts]

497. { Geometry of Environment }
S/A   [MARCH/STEADMAN;Methuen,1971(elementary)]

498. { Space Structures (general treatment)
S   [LOEB,A.L.(book,pp250)]

499. Formulation of finite-element algorithms
S/A   [BATHE/ODEN/WILSON,eds.,MIT Press,1977]

500. Mathematical developments arising from Hilbert's 23 Probs
A!   [AMS Proc.Sympos.Pure Math.No.28;pp628 (Inst.)]

501. Complexity of finite objects
S/@/A [ZVONKIN,A.K.,et al.,Russian Math.Surveys (1970)*]

502. Inversion of the LT without integration
S   [GOLDENBERG,SIAM Rev ?1965;WIDDER,'LT']

503. Inequalities among operators on probability distributions
S/A   [BECKENBACH,ed.,Genl.Ineqs.1133-149; MOYNIHAN,et al.]

504. Large-scale networks
S/A/A [BOESCH,ed.,IEEE Proc.1975]

505. Large deviations (for RVs)
S/A   [Ann.Prob.7(1979)745-789]

506. 'Bracing of rectangular networks'
δ/A   [SIAM J.Appl.Math.36,pp473-508;BOLKER,et al.]

507. {Algorithmic methods in probability(numerical)
δ/A   { [NEUTS,M.F.,ed.;North-Holland (Inst.)]

508. Electrical analogues in structural engineering
δ/A   [KEROPYAN/CHEGOLIN (ARNOLD,1967;book)]

509. Finite-element methods in stress analysis
δ/A   [HOLAND/BELL ,eds.,Proc.(?Trondheim,1969)]

510. 'HEX iff Brouwer fixed-point theorem'
δ   [AMM 86(1979)818-827]

511. Vector fields defined by complex functions
δ   [J.Diffl.Eqs.(1979)427-439;'Topol.Methods Fns CV(MORSE)]

512. Phase transitions as bifurcations
δ/A   [TAREEVA,Th.& Math.Phys.21,1189-1197;Symp.316,NY Acad Sci]

513. Iterative solutions for stochastic mappings
δ/A   [Proc.Sympos.U.Rochester,1976,343-370;MR 57 #15112]

514. Phase-space formulation of nonequilibrium/quantum SM
δ/@   [HOBSON,book;Physica,1970+,PJMB/TJS]

515. Symbol calculus(manipulations wit operators)
δ [Ann.Inst.H.Poincare 26A;J.Fnctl Anal.29;Ann. Phys'79+]

516. Global program optimization
δ/A/@ [SCHAEFFER,M.,Math.Th.Gl.Pr.Optn.,P.-H.,1973]

517. BV problems with discontinuous coefficients
δ [RASULOV,Methods of Contour Integration (N.-H.,1967)]

518. Algorithmic determination of Galois groups of alg.eqs
δ [Several implementations in SAM packages;SEE various Proc]

519. Singular perturbations for stochastic DEs
δ/A/@ [SIAM Rev.22(1980)]

520. Fuzzy Sets
δ [NEGOTIA,et al.(Wiley,1975,pp191);SEE BAMS]

521. Constructive approach to stochastic processes
δ/A [FREEDMAN,D.,Vols.1-3;Holden-Day (M.Ch;Brn Mtn;Approxn.]
/ papers* in refs. for Bishop/Bridges(bk)*

522. Possibilities for SAM in statistical mechanics
δ [de NEEF,J.Phys.A10(1977)801- ;also SW/W'loo,etc.]

523. Foundations of statl.mechanics ...
δ [KRYLOV,N.S.,Collected Works(USSR);?OUP/Princeton,1979]

524. Separation- of vars.VIA Lie grps vs (i)integration in
δ/A   finite terms (ii)factorization over alg.no.fields,
       (iii)solution of equations by radicals
       [SEE various SAM conf.proc.;ALSO J.Symbolic Computation]

525. Reconstruction of analytic functions given on small sets
δ   [PATIL,BAMS(1972) ... ;SW/W'loo]

526. RV/Integration;Spectral Synthesis (interconnections ...)
δ/A   [BENEDETTO;books (1976) ...]

527. Shape-dependent kernels for general domains
δ/A   [BESOV,O.V.,et al.,Integral Reprn of Fcns (Winston/Wiley)]

528. Theory of H^p-Functions
δ   [KOOSIS,P.,LMS Lect.Notes No.40;CUP;ALSO,Springer,c1990]

529. Algebraic geometry and symbolic integration
δ   [DAVENPORT,J.;LNCS No.102;ALSO:SAM Proc(var. authors)]

530. Methods/applications of interval analysis
δ/A   [MOORE,R.E.(SIAM,1979;book)]

531. { Nonnegative matrices in mathematical science }
δ/A   [BERMAN/PLEMMONS (AP,1979)]

532. { Problems and propositions in analysis }
δ/A   [KLAMBAUER;LNPAM No.49 (Dekker,1979)]

533. { Constructive linear algebra }
  δ   [GEWIRTZ,et al.(P.-H.,1974)   ]

534. Theory of best approximation/functional analysis
  δ/A [CBMS Regional Conf.Ser.13(1974);Bases in BSs(Singer)]

535. Analogue of Riemann surfaces for 2nd-order nonlin.oscilns.
  δ/A [Proc.4th Conf.Nonlinear Oscilns(Prague),p167(Abstract)]

536. { Discrete analytical mechanics }
  δ/@ [ROSENBERG,R.M.(book,1977)]

537. Generalized 'separation of variables'
  δ/@ [LNM 810 (KOORNWINDER);MILLER,W.(?AP)]

538. { Intro.to orthogonal polynomials }
  δ   [CHIAHARA (G&B,1978)]

539. { Qualitative analysis of large dynamical systems }
  δ/A [MICHEL/MILLER (AP,1977;pp289)]

540. { Monotone matrix functions and (D)AC }
  δ/A [DONOGHUE,W.F. (Springer,1974)]

541. {Newton polygon for algebraic/algebroid functions
  δ/A [VAINBERG-TRENOGIN(Noordhoff);ARTIN(G&B)]

542. CONVEXITY
  δ/A [Proc.AMS Sympos.Pure Math.VII (Inst.)]

543. Fast-Fourier-transform:pure or applied mathematics?
δ/@ [BAMS (11/79);HENRICI]

544. { Construction/applications of conformal maps }
δ   [BECKENBACH,ed.,NBS Sympos.,1952] /GAIER, D. (bks)

545. Finite-difference analogues of finite element schemes
δ/A [de BOOR,C.,Math.Aspect of FE (AP,1974;pp430)]

546. Approximation and abstract boundaries
A/@ [Amer.JM 85(1978)633-647;BAUER,H.]

547. Bifurcation for nondifferentiable operators
δ/@ [McLEOD/TURNER,ARMA 63(1976)1-45]

548. Sparse matrices(various aspects)
δ   [LNM 572]

549. Constrained nonlinear optimization
δ/A [BEIGHTLER,et al.,'Applied Geometric Progr.'(bk)*]

550. { Graphs and Hypergraphs }
δ   [BERGE,C.;Math.Library,No.6;(2nd ed.,1976)]

551. Nonlinear techniques in functional analysis
δ/@ [Nonlinearity and Functional Analaysis (AP,1977);Mel.B.]*'
     ALSO: HUGHES/MARSDEN (bk *')

552. Use of kernel functions in BV problems
δ/A [BERGMAN ,S.Numer.Mat.3(1961)209-]

553. { The kernel function and conformal mapping }
S   [AMS Surveys (2nd ed.;Inst..1st ed./SW]

554. Metrics on subspaces of BSs
S/A [Pacif.JM 13(1963)7-22;Ann.M.:67,517-573;53,250-]

555. Algebraic theory of integration methods(quadrature)
S/@ [BUTCHER,J.C.,Maths.Compn.(1972)] /later bk (↝ SIAM Rev.)

556. Boundaries of regions of(numerical) stability
S/A [STETTER(Springer);papers by BAKER,C.T.S.,and others]

557. Mathematical treatment of the 'renormalization group'
S   [Physics Reports 29(1977);BARBER]

558. CONFORM (SAM package for continuum mechanics)
S   [Proc.ASCE (1970)1239-1265]

559. { N-Dimensional Quasi-conformal Mapping }
S/@ [CARAMAN,P. (Abacus Press,1974)]

560. Unusual topics in 'calculus'
S   [CHAUNDY (OUP);Inst.]

561. PROXIMITY SPACES
S/A [NAMPALLY/WARRAK (CUP,?1974)]

562. Extremal positive-semi-definite forms(not reprbl as sqs)
S   [Math.Ann.231(1977)1-18]

563. Entropy and phase transitions in posets
    [J M Phys 8(1978)1711-]

564. Mean distance on a graph
    [Discrete Maths.17(1977)147-]

565. Deterministic/stochastic approximation of regions
    [DAVIS,P.J.;J.Approx.Th.21(1977)60-]
    et al.

566. { Applied Nonstandard Analysis }
    [DAVIS,M. (Wiley,1977)]

567. Inequalities for stochastic processes
    [DUBINS /SAVAGE (?SIAM/Dover;pp270)]

568. Introduction to 'C'(for SAM ...)
    [KERNIGHAN,et al.(?P.-H.)]

569. PROLOG (survey)
    [Logic for problem-solving (KOWALSKI;N-H/Elsevier,1979)]

570. Wide-ranging exposition of'Godel's theorem,etc.
    [Russ.Math.Surveys 29(1974)63-106]

571. Role of 'Newton polygon' in pesent-day mathematics
    [Transl.No.512C514 int.(Los Alamos Sci.Libr.Tchebotarev]

572. Acceleration of convergence in numerical analysis (BSs)
   A [LNM 584 ; BREZHINSKY(?)]

573. Geometric properties of spheres in NLSs
   δ [Duke MJ 52(1958)553-568]

574. (i) { Extension theory of topol.structures }
       [Sympos.,ed. FLACHSMEYER (1969)]
   δ/@ (ii) Extension of Analytic Objects
       [SIOU; Dekker(book)]

575. (i) Curve-Tracing
       [FROST,P. (Chelsea,1950)]
   δ (ii) A Book of Curves
       [LOCKWOOD,E.H. (CUP)]

576. Conformal mapping of nearly-circular domains
   δ/A [Pacific JM (1961)/bk (GAIER);KANTOROVICH/KRYLOV ,bk.]

577. Blended-function proc.for maps onto canonical domains
   δ/A [GORDON,J.E.,in AZIZ,Math.Foundations FEM (AP,1972)]

578. Changing the order of integrn in multiple singular integrals
   δ/@ [e.g.,GAKHOV ,BVP (Pergamon,1966)]

579. Approximation of functions
   A [GARABEDIAN,ed.,Sympos. (Elsevier,1965)]

580. Synthesis/analysis/formation of patterns
   δ/A [GRENANDER,U.;Lecs.in Pattern Th.vols 1-3 Springer,1976]

581. Notation for SHAPES of surfaces
    δ   [GRIFFITHS,H.B.;CUP,1976]

582. Polynomial approximation in plane elasticity
    δ/A   [Q J Mech. Appl.M IV(1951)444-448]

583. Completeness criteria for bases
    A/@   [SINGER,I.;Bases in BS vols I,II;MARTI (bks)]

584. Transmission of completeness from a CONS to nearby seqs.
    δ   [SINGER,I.;Bases in BS Vols I,II]

585. Role of COMPACTNESS in analysis
    δ/A   [AMM 67(1960) ; HEWITT,E.]

586. { Applied and Computational Complex Analysis }
    δ/A   [Vols.1(1973),2(1977),3(1986); HENRICI,P.]

587. Mechanizing Hypothesis-Formation
    δ/A   [LNM (1978);HAJEK/HAVRANEK ;ALSO: LNCS #32]*

588. Decomposition of large-scale problems
    δ/A   [HIMMELBLAU,D.,ed.(N.-H.,1973); SILJAK (N.-H.)]*,*

589. Approaches to the mathematical theory of switching circuits
    δ   [HU,S.-T.(U Calif.P,1968);MOISIL (Pergamon,1969)*]

590. { Topology and Maps }
    δ   [HUSAIN,T.(Plenum p,1977)]

591. { Innovative methods in engineering(numerical analysis) }
S/A [CETIM/Springer; Proc.,1977] *

592. Algorithms for approx. solution of singular integral eqs
S/A [IVANOV,V.V. (Noordhoff,1976)] *'–

593. Use of bi-polar co-ordinates in elasticity
S   [Phil.Trans.RS 221A(1921)265-290]

594. Diophantine representation of the set of prime numbers
S   [AMM (June/July,1976);JONES,et al.]

595. Extremal sets of convex sets
S/A [J.Fnctl Anal 26(1977)251-   ;JAYNE/ROGERS]

596. { Reviews of topics in numerical analysis }
S/A [JACOBS,D.A.N.,ed.,AP,1977;pp778]
S/A < ALSO: Acta Numerica : 1992 – (CUP) >

597. Abstract Analytic Number Theory
S/A/@ [KNOPFMACHER (N.-H.,1974) ;Inst.] *'

598. Linear operators and approximation
S/A [KOROVKIN,P.P. (G&B,1966)]

599. Continuity structures and spaces of mappings
S/A [Comm.Math.U.Carol.6(1965)257-278]

600. Construction of co-ordinates for separation of variables
S   [JMPhys.:14,1130-6;15,1025-32,1263-74,1728-37;16,499-501,

δ [512-7,2507-16;ALSO: 'Lie Groups and Seprn of vars']

601. Generalization of the 'Weierstrass density theorem'
A/@ [J.Approx.Th.21(11/77)]

602. Functions of a CV
δ/A [Sympos.,KAPLAN,W.,ed.;U. Mechigan P.,1955]

603. Use of splines in moment problems
δ/A [Chebychev Approximation (bk),KARLIN/STUDDEN]

604. Complexity of computation
δ/A [AMS/SIAM Sympos.(KARP,ed.);Sympos.,TRAUB,ed.(AP ?1976)]

605. Maximum-similitude subrelations (?substantial)
δ [KAUFMANN,Th.of Fuzzy Sets (AP,1975)]

606. { Functional equations for fns of one variable
δ { [KUCZMA,M.(PWN,1968);Inst.]

607. { The Markov moment problem }
δ [AMS transl.#50 (KREIN,et al.)]

608. { Uniform distribution of sequences }
δ/A [KUIPERS/NIEDEREITER (Wiley);BAMS (rev.)]

609. Integral representations of elastic potentials
δ [Potential Mehtods in Elasticity (KUPRADZE)]

610. Extension of 'Rouche's theorem' to functions of n CVs
δ/@ [LLOYD,N.G., JLMS (10/79)]

611. EUROSAM 79 (Various papers)
𝛿/A/@ [LNCS #72]

612. Papers on approximation theory/applications
A/@ [Approximations Theory Vols 1,2, ... (LORENTZ,G.G.,ed.)]

613. Reconstruction problems for TSs
𝛿   [LEADER,S.,in FLACHSMEYER,ed.(ext.th topol.struc.:#574]

614. Early LNM specially relevant for SiMA/MaAT
A!   [#:9,19,20,23,24,30,31,35,44,45,50,54,63,94,103,110,112,
       125,132,135,140,141,160,170,171,184,187,193,200,214,227,
       228,232,233,234,251,253,257,262,263,293,294,303,309,314,
       322,330 ...]

615. Threshold logic(switching theory)
𝛿   [LEWIS/COATES(1967);SHENG(1969);DERTOUZOS(MITP);MUROGA]

616. Probability theory of structural dynamics
𝛿/A/A [?CRANDALL,S.H.,et al. bks]

617. { Ergodic properties of algebraic number fields }
𝛿/A/@ [LINNIK,Yu.V. (?AP)]

618. Widths;entropy;reprn.by functions of fewer variables
𝛿/A   [LORENTZ,G.G.,Approximation Th.(HRWinston)]

619. Parallelism in numerical analysis (survey)
𝛿/A/@ [MIRANKER,SIAM Rev.13(1971)524-547]

620. Calculus of functions of n noncommuting variables
S/@ [MASLOV,'Operational Methods' (MIR ?1978)]

621. Orderable groups
S/@ [BOTTO-MURA,et al.,eds.(?),LNPAM (Dekker,1977)]

622. Dirichlet series(general theorems/properties)
S/A [MANDELBROJT,S. (Reidel,?1972)]

623. Algebraic Theories (wide scope/historical aspects)
S/@ [GTM #26 ; MANES (Springer,1976)]

624. Topologies on spaces of subsets
S/@ [TAMS 71(1951)152-182]

625. Weak separation of variables
S/A/@ [MOON/SPENCER,Field Theory Handbook (Springer,1961)*]

626. Quantum mechanics as a statistical theory
S/A [MOYAL,J.E.,PCPS 45(1949)*]

627. Prime factors and additive functions (Bibliography)
S [NORTON,K.K. (1979+,Boulder Col.) ?Acta Math.]

628. Effectiveness of seives
S/A [NEWMAN,D.J./RICHARDS,I.;J.Number Th.10(4),1978]

629. Reprns.of conformal maps onto canonical domains
S/A [NEHARI,Z.;DUKE MJ 16(1949)165-178]/ Nehari = bk (Dover)*

630. n-variable approximations and embedding theorems
δ/A [NIKOL'SKII (Springer,1975)]

631. Iterative solution of n-variable algebraic equations
δ/A/@ [ORTEGA/RHEINBOLDT; (AP,1970)]

632. The Newton boundary (bifurcation theory)
δ/A/@ [J.M.Soc.Japan (7/79)]

633. Approximately differentiable functions ('r-topology')
δ/A/@ [Pacific JM 72(1978)207-222;SAKS (Int.);MANN (Addn Thms)]

634. Reconstruction of analytic functions
δ [PATIL (BAMS);Adv.Prob.(YOUNG,L.C.)]

635. Use of analytic functions in bi-harmonic analysis
δ [TAMS 59(2)(1946)248-276;ALSO:books on theoretical elast.]

636. Network analogues of vector analysis/thermodynamics/...
δ/A [MRI Sympos.Proc.XVI,1966;ALSO:other MRI Symposia]

637. h as a function of f (for f(hx)) in Taylor's thm
δ/A [PRASAD,G.'Six Lecs.on Taylor's Thm'*] *Inverse problems!*

638. Algebraic properties of Elementary functions
δ/@ [RISCH,Amer.JM101(10/79)743-759;BASS,et
    al.,eds.,Contrib.Alg.(1977)329-342;
    PAMS (1976)1-7;Ann Mat.93,252-268;TAMS 27,68-90;
    Pacif.JM 59,535-]

639. Co-ord.systems as reprns (of objects/motions/...)
S   [Refs.to be supplied(e.g.,GRG)]

640. Confinement regions for collections of charged particles
S/A/A [CHAPMAN,S./BARTOLDS,VolII;books on plasma theory]

641. No.of solns.of discrete Theodorsen equation(conformal maps)
S   [Numer.Mat.39(2); Math.Compn 31,478- ;Numer.Mat.36,405-]

642. Branching of solutions of operator equations (survey)
S/A   [Rocky Mountain JM (1973)203-250 ;?STEWART,J.]

643. { Multiple-Conclusion Logic(graphical notation) }
S   [SHOESMITH/SMILEY;CUP,1978   ]

644. Construction of +ve polyn not reprbl as sum-of-squares
S   [Math.Nachr.88(1979)385-390]

645. Kinematic measures for collisions of convex bodies
S/A   [Mathematika (6/78)]

646. Survey/(computer-)solution of 4-Colour Problem
S   [SAATY/KAINEN,bk; Inst.;Grundl.#209(RINGEL/YOUNGS)]

647. Evalution of integrals VIA geometric invariants
S   [SANTALO,Integral Geometry(A-W/CUP,pp450)]

648. Constructive/optimal proc.in linear approximation
S/A   [SARD,A.,Linear Approx.,AMS Surveys;pp544 *-]

649. Global program optimization(graph-theoretical)
  δ/A/@ [SCHAEFFER,M.(P-H,1973);later book*]

650. Geometry of spheres in NLSs(inner metric geometry)
  δ [LNPAM #20 (Dekker,1976),pp256;SCHAFFER]

651. General approximation problems involving SPLINES
  A [SCHOENBERG,I.J.,ed.,Sympos.(AP,1969);SCHUMAKER (bk)]

652. What is Distance? (elem.but ingenious)
  δ [SCHREIDER,Yu.,U.Chicago P]

653. Multi-valued Legendre transformations(nonlin.elast./optzn)
  δ/@ [MPCPS 82(1977)147-163/bk(c1988);SEWELL,M.J.]

654. Structure and approximation in physical theory
  δ/A/A [HARTKAMPER,A,et al.,eds.(Plenum P,1981;pp264) *-]

655. Construction of conformal maps of 2-connected domains
  δ [Devel.Th.&Appl.Mech.Vol.3 (SHAW,ed.)]

656. Best approx.in NLSs by sets of elements from subspaces
  δ/A [SINGER,I.(Springer,1970);Inst.]

657. Bases in Banach Spaces
  δ/A [SINGER,I.,Vols.1,2;Springer,1970,1982]

658. { Encyclopedic Dictionary of Engineering Mathematics }
[SNEDDON,I.N.,ed. (Pergamon,1976;pp808]

659. { The Hypercircle in Mathematical Physics (Pre-FEM) }
δ/A/@ [SYNGE,J.L.;CUP,1957,pp424]

660. Regularization for ill-posed problems(PDEs,etc.)
δ/A/@ [TIKHONOV/ARSENIN (Winston/Wiley,1977)*]

661. '100 Years of the Plane Problem of Elasticity'
δ [Appl.Mech.Revs.17(1964)175-186 (TEODORESCU,P.P.)]

662. '100 Years of Mathematics'(broad survey,P/A)
δ [TEMPLE,G.;DUCKworth,1980)]

663. Arbitrary curvilinear co-ordinates in planar elasticity
δ [Rev.Mech.Appliquee(Roumania)8:453-479;589-609;953-969]

664. Symposia on general topology
δ/A/@ [Prague;several vols.]

665. Symposia/conferences on nonlinear oscillations
δ/A [Prague Symposia,Nth,...]

666. { Combinatorial Optimization (broad;algorithmic) }
δ/A [PAPDIMITRIOU/STEIGLITZ (P-H,1982)* pp496]

667. { Poential theory in modern function theory }
δ/A/@ [TSUJI,M. (Chelsea,1975;pp590) Inst.]

668. Various applications of set theory
   δ   [Van DALEN,D.,et al.(Pergamon);last Ch.*]

669. Generalized analytic functions
   δ/@   [VEKUA,I.N. (Pergamon,1962)*;GILBERT/BUC'N(AP,1983)*]

670. Lattice states of minimal potential energy(1-D)
   δ/A/A [VENTEVOGEL ;Physica (7/78,and later)*-]

671. Functional analysis in partially ordered spaces
   δ/@   [VULIKH (Noordhoff,1967);KANTOROVICH,et al. (??)]

672. { Fixed points and almost-fixed points
   δ/A   [Van de WALT (Math.CentrumTract #1 (Amsterdam,1963)]

673. { Extensions and embeddings in analysis (manifolds)
   δ/A   [WELLS/WILLIAMS (Ergebnisse;Springer,1975)]

674. Finite elements and algebraic geometry
   δ/A/@ [WACHPRESS,E.'A Rational FE Basis'(Wiley,1975)]

675. { Mathematics of finite elements and applications }
   δ/A   [Vols I,II,III,IV,V,VI,... ;WHITEMAN,J.R.,ed.(?AP)]

676. Hypercomplex Laplace and Fourier transforms
   δ/@   [SOMMEN,F.(prepr.*);BELTRAMI/WOHLERS (bk);Inst.]

677. Runge's theorem in hypercomplex function theory
   δ/@   [DELANGHE,et al.,J.Approx.Th.(1980+)]

678. Algebraic/topological characterizations of ANALYTICITY
S/@ [Adv.Math.5(1970)311-338 (RICHARDS,J.)]

679. Irrationality of linear combns of radicals(VIA Galois th.)]
S/A [Adv.Math.13,268- *] (Richards, J.I)

680. Asymptotics for block-Toeplitz matrices
S/A/@ [Adv.Math.13,284-322 (WIDOM,H.)]

681. Discrete convergence/stability (for differentiable maps)
S/A [Rev.Roum.Mat.PA(1976)*;STUMMEL,F. (p)repr.*]

682. Constructive functional analysis
S/A [Pitman Res.Notes #28 (BRIDGES,D.)]

683. Foundations of constructive analysis
S/A/@ [BISHOP,E.(McG-H,1967)*,2nd ed./Bridges(Springer,Grundl)]

684. Various ingenious applications of math.procedures
S/A/@ [BAMS 80,1053-1070* (DUFFIN,R.J.)]

685. Constructive solution of'potential scattering'
S [PEARSON,D.B.(prepr.)* /?Comm.Math.Phys.]

686. Converse of Lebesgue covering property
S [Bull.LMS 4(1972)184-186*]

687. A new transcendental function
S [Bull.LMS 4,167-170*]

688. Involutory functions and nets(algebraic formulations)
δ/@   [Stud.U.Babes-Bolyai (1974);ACZEL/RADO]

689. Algebraic K-theory (intro.)
δ   [AMM 82,329-364 (LAM,et al.)]

690. Finite projective planes/block designs
δ   [MANN,H.B.(Dover);HIRSCHFELD (OUP)]

691. Bibliography on computers in group theory,etc.
δ   [FELSCH,V. (Aachen)*;c1978]

692. Algebraic formulations of 'language theory'
δ/@/A [Bull.LMS 7,1-29 (COHN,P.M.)]

693. Riemann hypothesis(RH)as a nonlinear optimization problem
δ/A   [Mathematika 22(1975)92-96]

694. Extremal problems for 'posynomials'
A/@   [DUFFIN,R.J.,et al.,Geometric Programming (Wiley,1967)*]

695. Nonlinear almost-periodic oscillations
δ/A/@ [KRASNOSEL'SKII,et al.(Halstead/Wiley,1973;pp326)*]

696. Sparse network equations
δ/@   [BRAMELLER,et al.,Sparsity (Pitman,1976)*]

697. Geometry of convex figures (expostion/problems)
δ [YAGLOM,I.M.,et al.(Holt-R-W,1961;pp301)*]

698. Geometry as a 'board game'
δ/A [GOODSTEIN/PRIMROSE (U.of Leicester,1962)*]

699. Simple treatment of 'reliability'
δ [KAUFFMAN,A. (Transworld)*]

700. Representations of molecular vibrations
δ [GANS,P. (C&H,1971;pp236),esp.,Ch.4]

701. Grammar of Dirac Matrices (Clifford algebras,etc.)
δ/@ [RAMAKRISHNAN,A.(Tata/McG-H,1972)*]

702. Error-propagation for finite-difference methods
δ/A [HENRICI (?Wiley)* ;WILKINSON,J.M.,books(HMSO/OUP)]

703. Abstraction and analogy
@/A [Abstraction mappings... (paper in Sympos.(?LNCS)*]

704. Techniques to increase parallelism/efficiency of algorithms
δ/A [Church-Rosser procedures... (paper)*]

705. Factorization of n-var.polynomials over algebraic no.fields
δ/@ [Math.Computation 30(1976)324-336 (WANG,P.S.)*]

706. Convergence in probability and Lambda-metrics
δ/A [LUKACS,E.,Stochastic Convergence (Heath Math.Mono.1968)*]

707. Use of the Fast-Fourier Transform in complex analysis
δ   [SIAM Rev.21(1979);HENRICI*]

708. Bracing of rectangular networks
δ/A [SIAM J.Appl.Math.36(1979)473-508]

709. Stochastic processes and Special functions
δ   [J.Math.Anal.Appl.61(1977)262-291(HOARE,M.R.,et al.)*]

710. Development of Riemannian geom.VIA volumes of small balls
δ/@ [Acta Math.142,157- *]

711. Stability domains for numerical solution of integral eqs
δ/A [BAKER,C.T.H./Keech*;BAKER(OUP)*]

712. Theorems on zero-distributions for seqs of analytic fns.
δ/A [Ark.f.Mat.3(1)(?1953)1-50 (GANELIUS)*;ROSENBLOOM,P.C.*]

713. Metrical properties of 'block'and 'operator' matrices
δ/A/@ [OSTROWSKI,A.,J.Math.Anal.Appl.2,161-209*]

714. Various general results in complex approximation theory
A   [GAIER,D.(book),1982+]

715. Connections between LOGIC and PROBABILITY
δ/A [FENSTAD,J.E. (Prepr.,Oslo)*]

716. Nonstandard analysis for stochastic processes,etc.
δ/A [FENSTAD (prepr.,Oslo)*]

717. Brownian motion and classical analysis
δ/A [Bull.LMS 8,145-155*]

718. Analytic capacity of sets in approximation theory
δ/A [VITUSHKIN ;Russ.Math.Surveys(1967+)*]

719. Factorization of operators
δ/@ [J.Fnl.Anal.9,262-295*;JDE 14,518- * (SCHUMITSKY/McNABB)]

720. Plane-filling regular curve families
δ [I,II;Duke MJ (1940)* (KAPLAN,W.)]

721. { Geometric Functional Analysis }
δ/A [HOLMES,R.B. (Springer)*]

722. Quasi-isomorphisms (cybernetics)
δ/A [GHOSAL,A.(G&B,1978);p81]

723. Constructive aspects of bifurcation
δ/A [in: Sympos.,RABINOWITZ,P.,ed.,pp1-35*]

724. Generic bifurcation
δ/A [HALE,J.K.,in:Sympos.Nonlin.Anal.#1(Heriot-W);KNOPS,ed.]

725. Imperfect bifurcations treated VIA singularity theory
δ/A/@ [Comm.P&A M (1979)21-98*]

726. Constructive techniques for solution of branching equations
δ/A [DEKKER, et al., in: Sympos.(HOLMES,P.J.,ed.,1980)1-17*]

727. Computation of invariant manifolds
δ   [Sympos.(HOLMES,P.J.,ed.,1980)27-42*]

728. Equilibrium states /convexity in T-D/stat.mech.
δ   [WIGHTMAN,J.R.,Intro.to book(lattice gases)by ISRAEL,R.B.]

729. Completely-additive functions and dynamical systems
δ   [BIRKHOFF/SMITH,Surface Trans.*;other BIRKHOFF papers*]

730. Phase transitions and bifurcation
δ/A [Ann.NY Acad.Sci.316,417-432(?ROSS)*]

731. Reduction of CALCULUS to ALGEBRA
δ/A [KOCK,A.,Synthetic Differential Geometry,CUP(LMS)#51*]

732. Algebraic structure of queuing processes
δ/A [KINGMAN,J.F.C. (Methuen Rev.Ser.#6,pp44)*]

733. { Topology from the differentiable viewpoint }
δ/A [MILNOR,J.,bk*]

734. Stabilizability for linear PDEs
δ/A [SIAM Rev.20,639-739 (?RUSSELL)]

735. Problems of integral geometry
δ   [ROMANOV (Springer)*-]

736. Miscellaneous papers on symbolic computation
δ   [Proc.EUROCAM 82 (LNCS 1982+)]

737. Flow in transportation networks
δ/A   [NEWELL,G.F.(MIT P,1980)]

738. Smoothing and approximation of functions
δ/A   [v Nostrand Math.Stud.#24;SHAPIRO,H.S.]

739. Euclidean/noneuclidean metrics in function theory
δ/A/@ [NEVANLINNA,R.,Analytic Functions (Springer)*-]

740. Point sets of harmonic measure zero
δ/@   [NEVANLINNA,Anal.Fns.,Ch.5*]

741. Nonlinear Hardy spaces and electrical power transfer
δ/A   [BORDER,J.,PhD thesis,U Calif San Diego,1979]

742. Extendability of proper holomorphic mappings
δ/@   [BAMS (7/82)265-272*]

743. Picard's 'little thm' for Riemann surfaces/Brownian motion
δ/@   [BAMS (7/82)259-263*]

744. Fast recursion formulae for reprns of Lie algebras
δ/A   [BAMS (7/82)237-242*]

745. Invertible maps between Frechet spaces
   δ   [BAMS (7/82)65-236*]

746. Noneuclidean functional analysis and electronics
   S/@/A   [BAMS (7/82)1-64*]

747. Generalized contraction mapping principle
   δ/@   [KRASNOSEL'SKII,et al.,Nonlin.AP Oscl.*;p132]

748. Approximation problems in mathematical statistics
   S/A   [LNCS #32,pp258-266*]

749. 'Logics of discovery'
   δ/A   [LNCS #32,pp30-46*]

750. Information algebras(over networks)
   S/A   [Proc.MRI Sympos.(1962),Math.Th.Automata*415-435]

751. Algebraic synthesis of derivation of the FFT
   S/@   [Proc.Sympos.MRI (1971)*pp359-377]

752. ? Risch-type algorithms for integration over groups
   δ/@   [BAMS,1970;TAMS,1969; ...]

753. Hyperbolic geometry and approximation on finite sets
   δ/A   [RICE,J.R.,Approximation,Vol 2*]

754. Asymptotic inversion of convolution tranforms
   δ/A   [WIDOM,H.(prepr.)*]

755. { Information measures:characterization/generalization }
δ/@ [ACZEL,J.,et al.,eds.,Sympos. (AP)*]

756. Use of linear information for solution of nonlinear eqs.
δ/A [WASILKOWSKI*;TRAUB(/et al.)*]

757. Iteration for polynomial eqs with linear information
δ/A [Report CMU #79-138*]

758. Structure of parallel algorithms
δ [CMU Rep.(KUNG) 79-143*]

759. Reconstruction of finite automata
δ [Rep.CMU #82-127* (Sec.4)]

760. Analogical problem solving
δ/A [CMU #82-126*]

761. Complexity of composition of power-series
δ [CMU #78-128]

762. Parallel algorithms(construction of minimum-spanning trees)
δ [Rep.CMU (9/79);BENTLEY,J.L.*]

763. Automatic search procedures(for VLSI machines)
δ [BENTLEY/KUNG (CMU)*]

764. Area-time complexity for n-bit binary multiplication
δ/A [BRENT/KUNG ;CMU(1979)* { 'Uncertainty principle'}]

765. Approximate assertions(in program analysis)
S/A/A [SHAW,M. (CMU,1979)*]

766. Possibility of SYSTOLIC optimal matrix multiplication
S/A [CMU #79-103*]

767. Regular layouts for parallel adders
S [CMU #79-131*]

768. Some basic algorithms for VLSI systems
S [CMU 79-151* ;ULLMAN (bk)*]/KORTE, B. +(LNM #190)
 SHERWANI (Kluwer)*

769. Formal calculatins for stochastic DEs
S [LAMNABHI (Thesis(French))*]

770. Symbolic computation with noncommutative operators
S/@ [LAMNABHI (thesis)*]

771. MACSYMA computations for anisotropic shallow shells
S [ANDERSEN/NOOR (NASA Report)*]

772. Predicate-path expressions AS nondeterministic programs
S/A [CMU #79-134*]

773. Optimal error algorithms/complexity
S/A [TRAUB/WOZNIAKOWSKI (ACM,1980)*]

774. Some basic SAM algorithms
S/A [GEDDES,K.O./R.MOENCK (draft Chs for bk)*]
→ bk: G/C/L 'Alg^{ms} Comp.Alg. (Kluwer, 1992)*

775. Algebra/algebraic computing(basic theory)
δ [LIPSON (A-W,1981)*]

776. Basic calculational scheme for (SM) renormalization
δ/A [WILSON,K.G. (Cargese LN)*]

777. 'Stochastic behaviour'of the Zeta function' ALSO: BERRY, M.C. +
$\delta/A/@$ [KLEIN,C. (??BAMS/Mathematika(UCL))]   PRSL (A) *

778. Difficulties with 'homotopy proofs of fp-theorms'
δ [?BACHEM,et al.,eds.,Mathl Progr ... (Springer,1983)*]

δ 779. (i) FS-reprns for periodic C^2 - functions
δ/A (ii) Max.principles for classes of complex fns.
       [e.g.,ZYGMUND,A.,Trig.Series (CUP,1959)*]

780. Integral equations whose kernels are Szego kernel-functions
δ [PLMS #39(1960)376-394]

781. Elimination/cancellation of critical points of Morse fns
δ/A [MILNOR,J.,Lecs.on h-cobordism theorem (bk)*]

782. Justification of higher-dimensional network analogies
$\delta/@/A$ [Proc.Sympos.Large Engrng Systems II(W'loo)*357-363;
     KIM/CHIEN:Topol.Anal.&Synthesis/N-W (bk)* ;MRI
     Sympos.(1966)* 453-491]

783. Method of summary representation(Numer.anal./PDEs)
δ/A [POLOZII (pergamon,1965)]

784. Topological descriptions of molecular structure
δ/A [Proc.Nat.Acad.Sci.74(1977)2616-2619* ;bk (Elsevier)*]

785. Reconstructions of solutions of diffusion equations
δ [Dokl.17(1976)914-916*]

786. Construction of Riemann surfaces with prescribed branch-pts
δ [SIBUYA (bk;N-H); Inst.]

787. Derivation of Poincare maps for forced ODEs
δ [BERNUSSON,J.(Pergamon,1966)]

788. Determination of domains of attraction for 2nd-order ODEs
δ [Rend.Mat.(6)10(?1977)417-431*?]

789. Convergence thms for 1st order ODE systmes with dissipation
δ/A [Diffl.Eqs.(USSR)13(1977)964-972*?;MR 57 #10086(Lienard)]

790. 'Resultants' for pairs of analytic functions
δ/@ [Applic.Anal.7(1977/78)191-205;MR 58 #2422]

791. Papers on the theory of parallel programming
δ/@ [Adv.Inf.Syst.Sci.6,1-55;6,57-113]

792. Entropy/phase transitions in posets
δ/A [J.Math.Phys.19(1978)1711-1713*? ; MR 58 #20062]

793. Matrices with some prescribed elements
δ [Israel JM 11(1972)184-189; MR 52 #431]

794. A 'Rouche theorem' for functions of n CVs
δ/@ [JLMS 20(1979)259-272; MR 52 #431*]

795. Automatic translation of seql.progr.to parallel progr.
δ [Cybernetics(USSR) 10(1974):1-18;197-212]

796. Representation of n-var.fns in terms of fns of fewer vars
δ [Proc.Int.Conf.Approx.Th.(U M'lnd,1970); MR 50 #845;
    J.Approx.Th.6(1972)123-134]

797. Analogue of Floquet theory for Jacobi matrices
δ/A/A [Invent.Mat.37(1976)45-81; MR 58 #31266]

798. Heuristics of the 4-Colour- Problem proof
δ [J.Graph Theory 1(1977)193-206; MR 58 #27600]

799. Cohomology in physics/q-connectivity in sociology (???)
δ/A [Int.J.Man/Machine Studies 4(1972)139-167* (ATKIN,R.H.)]

800. Prime numbers and Brownian motion
δ/A [AMM 80(1972) 1099-1115* ;MR 49 #9883]

801. Representation theorems for distribution functions
δ [PLMS 8(1958)177-223*]

802. Nevanlinna theory and Brownian motion
δ/A/A [CARNE,K.Prepr.,Cambridge (?)]

803. Generalized subadditive functions
δ/A/@ [SHISHA,O.,ed.,Inequalities (AP,1967)127-135]

804. Lambda-calculus basis for SAM
δ [Indag.Mat.34(1972)381-392*? ; MR 48 #71]

805. Replacement of continuous bodies by frames(?F-E)
δ/A [MR 55 #7070]

806. Restricted infinite sum of {Q-form(5 vars)}$^{-s}$ as Zeta fns
δ [J.Math.Phys.16(1975)1237-1238; MR 55 #7228]

807. Extensions of topological spaces
δ/@ [LNPAM 24(Dekker,1976)129-184; MR 55 #6368]

808. Metrization of the hyperspace of a MS
δ/A [Fund.Mat.94(1977)191-207; MR 55 #6373]

809. Gronwall-type inequalities
δ/A [Carleton Math.Lec.Notes #11(1975,pp123)]

810. (i) Singularities in solutions of PDEs
δ [HAWKING/ELLIS (bk,CUP)* ;ALSO:in elasticity theory ...]
(ii) Ineqs.for difference,differential,and integral eqs.
δ/A [Ind J PA M 6(1975)1479-1487;Proc.Nat.Ac.Sci.Ind.A46,21-26]

811. 'Coincidences' and fixed-point theorems
δ/A [Fund.Mat.90(1975)131-142;ALSO: MR 53 #4027]

812. Computational reducibility
   δ   [J.Comp.Syst.Sci.12(1976)122-131;ALSO: MR 53 #4616]

813. Parallel computation in linear algebra(Bibliography)
   δ/@   [FADEEV(A) (1975); MR 53 #4500]

814. U.cgce of 'discrete',to'cts,Green's fns implies'stability'
   δ/A   [Applbl Anal 14(1982)73-98 (BEYN)]

815. Generalizations of Fatou's inequality
   δ/A   [BELLOW,S.,et al.;Ann Inst H.Poincare18(1982)73-98]

816. Spectral estimation
   δ/A   [J.Math.Anal.Appl.91(1983)444-509]

817. Differentials of fuzzy functions
   δ/@   [J.Math.Anal.Appl.91(1983)522-528]

818. Approximation of roots of f(x,y)=0 (f,polyn.) {alg.geom.}
   δ/A/@ [J M Soc.Japan 34(1982)637-652]

819. Best diophantine approximation to sets of linear forms
   A   [J Austral.M S A34(1983)114-122]

820. Relative distance (for error analysis)
   A   [Maths.Computatn. 39(1982)563-569]

821. Summability and (D)AC
   δ/A   [TOMM; J fd R&A M 339(1983)133-146]

822. Distance between two vectors with given dispersion matrices
δ/A [Lin.Alg.Applns.48(1982)257-263]

823. Scaling(of matrices)to doubly stochastic form
δ [Lin.Alg.Applns. 48(1982)53-79 (PARLETT,B.N.,et al.)]

824. Estimates for multi-variable trigonometrical sums
δ/A [Proc.Steklov Inst.Math. (1982) #2]

825. Linear characterizations of optimization problems
δ/A [SIAM J Computing 11(1982)620-632]

826. Area/time trade-offs for VLSI
δ/A [SIAM J Comp.11(1982)737-747]

827. Category-theoretic solution of recursive domain equations
δ/@ [SIAM J Comp.11(1982)761-783]

828. Probabilistic AP functions
δ/A/@ [Rend.Mat.(1982)613-626]

829. Perturbation expansions over perturbed domains
δ/A [SIAM Rev.24(1982)381-400]

830. Sequence transformations
δ [WIMP,J.(AP,1981,pp257)]

831. Theory of unimolecular reactions
δ/A [SLATER,N.B.(bk,Cornell UP)* ;ALSO:NIKITIN(OUP)*]

832. Generalized 'Lorenz' PDE system(strange attractors?)
δ/A [Comm.Math.Phys.60(1978)193-204 ; MR 58 #13804]

833. Analytic Inequalities(general/specific)
δ/A [MITRINOVIC,D.S.(Springer)* ;MR 43 #448]

834. The 'part-metric' in convex sets
δ/A [Pacif.J M 30(1969)15-33;MR 43 #859]

835. AP functions and functional equations
δ/A [AMERIO/PROUSE (v Nostrand,1971;pp184); MR 43 #819]

836. Center-manifold theory (reduction)
δ [CARR,J. (Springer)*]

837. Geometric mean of convex sets
δ/A [MR 51 #6591* (?Steklov/Leningrad)]

838. Everywhere-differentiable fns and the density- topology
δ/A [PAMS 51(1975)250 ; MR 51 #3366]

839. Invariant subspaces(survey)
δ [RADJAVI,et al.(Springer,Ergebn.Bd 77(1973);MR 51#3924+]

840. Rearrangement-invariant norms
δ [?? refs. (bks on fnl anal/oprs.(?ZYGMUND)]

841. Almost-convergence
S/A [JLMS 17(1978)317-320; MR 58 #1820]

842. Summability in abstract structures
S/A [MR 58 #1824; MR51 #13516]

843. Functional inequalities
S/A [MR 58 #1795]

844. Asymptotic representation of functions ...
S/A [HENRICI 'CV'(Wiley)*;OLVER (AP)*- ; SIROVICH (Springer)*]

845. Simultaneous contractification
S/A [J.Res.NBS 73B(1969) (:=NBS73B) 301-305)*]

846. Nonlinear trans. for convergence acceleration of sequences
S/A [NBS73B: 25-29;251-273]

847. Minimax error-selection/adjustment
S/A [NBS73B: 225-230;231-239]

848. Automorphic integrals with pre-assigned periods
S [NBS73B153-161]

849. Cuttable and cut-reducible matrices
S [NBS73B 145-151]

850. Sufficient conds.for instability of quadrature methods
S/A [NBS73 119-123]

851. Traffic assignment in generalized networks
δ/@ [NBS73 91-118]

852. Effective determination of $f(x)$ from subseq.of $\{f_n(0)\}$
δ [C-R Ac.Sci.Paris AB281(1975)551-554; MR 52 #5918]

853. 'Integration in finite terms' for line- integrals(n vars.)
δ/@ [Comm.Alg.3(1975)781-795; MR 52 #5637]

854. An ineq.valid for'spherical',but false for 'convex'measure
δ/A [PCPS 60(1964)821-845*]

855. Chebychev approx.of one rational fn.by another(estimates)
A [PCPS: 60(1964)877-890* ; 58(?1962)244-267*]

856. Extensions of Liapunov's majorization principle
δ/A/@ [PCPS 60,891-896]

857. Random differential inequalities
δ/A/@ [LAKSMIKANTHAM/LADDE (bk,AP,1980);proof copy*]

858. Rearrangeable connecting networks
δ/A [PCPS 60,939-948]

859. Nonexpansive mappings and fp-free isometries
δ [PCPS 60,439-447]

860. L/R densities on arcs of positive 2-D measure
   δ   [PCPS 60,517-524]

861. Central Limit Theorem ... for finite Markov chains
   δ/A [PCPS 60,547-567]

862. 'Phase-space formulation' of quantum mechnaics
   δ/@ [PCPS 60,581-586;cf.,Phys.Rev.40(1932)749-759(WIGNER)]

863. Normal modes in nonlinear systems
   δ/@ [PCPS 60,595-611]

864. Additive functionals on groups
   δ/@ [PCPS 58,196-205]

865. Measures of asymmetry of convex bodies
   δ/A [PCPS 58,217-220]

866. Bounds on zeros of general polynomials
   δ/A [PCPS 58,229-234]

867. 'Factorizable matrices' for finite Markov chains
   δ   [PCPS 58:268-285;286-298]

868. 'Topological degree' for noncompact mappings
   δ   [PCPS 63,335-347]

869. Space-times with two partial orderings
   δ/A [PCPS 63,481-501]

870. Hausdorff measure of planar Brownian paths
δ/A [PCPS 57(1959) ...]

871. Hausdorff dimension of general Cantor sets
δ   [PCPS 61,679-696]

872. Well-quasi-ordering of infinite trees
δ/@ [PCPS 61,697-720]

873. Smoothness/singularities of statistical distribution fns.
δ   [PCPS 61,721-739]

874. Infinite systems of linear algebraic equations
δ/A [PCPS 61,781-794]

875. Cohomology operations and duality
δ   [PCPS 64,15-30]

876. Compatible uniformities
δ   [PCPS 64,37-40]

877. Representation thms. for partially-ordered Banach algebras
δ/A/@ [PCPS 64,53-60 ; ALSO: WONG/ING (bk,OUP)*,Ordered TVSs]

878. Density of '2-D Jordan curves' in the plane
δ/A [PCPS 64,67-70]

879. Uniqueness theorems for n-variable analytic functions
δ/@ [PCPS 64,71-82]

880. Uniform approximation by Fourier-integral transforms
S/A   [PCPS 64: 323-333;615-623]

881. Paranormed sequence spaces generated by infinite matrices
S/A/@ [PCPS 64,335-340]

882. Sequential convergence in l.c.Hausdorff spaces
S/A   [PCPS 64,341-363]

883. Stability of 'wedges' and 'semi-algebras'
S/A/@ [PCPS 64,365-376]

884. Generalization of the Stieltjes transform
S/@   [PCPS 64,407-412]

885. Generalization of Laplace-Hankel transforms
S/@   [PCPS 64,399-406]

886. Lower bounds for the n-dimensional 'dimer problem'
S/A   [PCPS 64,455-463]

887. Convergence rates for the Law of Large Numbers
S/A   [PCPS 64,485-488]

888. Construction of 3-manifolds from 2-complexes
S   [PCPS 64,603-613]

889. Stable subspaces of the direct sum of L(1) and L(inf.)
   [PCPS 64,625-643]

890. Construction of an inner measure on a von Neumann algebra
   [PCPS 64,645-650]

891. Expansion in series of Gontcharoff polynomials
   [PCPS 64,699-703]

892. Characterization of 'controls'
   [PCPS 64,741-748]

893. Comparison principles in stability theory
   [PCPS 64,749-756]

894. Inequalities for functions regular in the Unit Disc
   [PCPS 58,26-37]

895. 'Sequential Montecarlo' (for evaluation of integrals)
   [PCPS 58,57-78]

896. Proof of Macmohan's 'Master Theorem' VIA n-CV integration
   [PCPS 58,160-]

897. Convexity ... in preference analysis
   [PCPS 59,287-305]

898. Partial ordering on linear TSs
   [PCPS 59,323-327]

899. Co-positive, and completely-positive, quadratic forms
δ/A [PCPS: 59,329-339;58,17-25;JLMS 36(1961)185-192]

900. Inequalities for weighted arithmetic and geometric means
δ/A [PCPS 59,341-346]

901. Dual-integral-equations for Bessel-function kernels
δ [PCPS 59,351-362]

902. Dual-series-equations involving Jacobi polynomials
δ [PCPS 59,363-371]

903. Liapunov functions for ODEs over BSs
δ/@ [PCPS 59,373-381]

904. Cgce of sums of IRVs with singular covariance matrices
δ/A/@ [PCPS 59,411-416]

905. Invariant stochastic processes and group representations
δ/@/A [PCPS 59,431-450]

906. Weakly stochastic ODEs
δ/@ [PCPS 59,463-481]

907. Extension of the Carleman(CV) formulae to subharmonic fns.
δ/@ [PCPS 62,51-59]

908. An interpretation of 'negative probabilities'
δ/@ [PCPS 62,83-86]

909. TSs and their lattices of open subsets
S/@ [PCPS 60,197-207]

910. Order-continuous measures and Baire category
S/A [PCPS 60,205-207]

911. Efficiency of Montecarlo integration
S/A [PCPS 60,357-358]

912. Martingale inequalities for queues
S/A [PCPS 60,359-361]

913. The use of algebraic structures in physics
S/A [PCPS 57,851-864]

914. Approximate continuity
S/A/@ [e.g.,GOFFMAN,C.(bk,Rinehart,1953,189,ff]

915. Approximate uniform convergence
S/A/@ [GOFFMAN (bk)186,ff]

916. Applications of well-ordering
S/A [GOFFMAN (bk)146-152]

917. Approximation of bi-variate fns.by univariate fns.
A [Numer.Mat.39(1982)65-84]

918. Bounds for condition-numbers of matrices
δ/A   [Numer.Mat.3985-96]

919. Effective procedures for algebraic integrability of flows
δ/A   [Invent.Mat. 67,297-331*]

920. 'Vanishing theorems' for power-series
δ   [Invent.Mat.67,275-296*]

921. Invariants of 1-D characteristics and integration of PDEs
δ   [PLMS11(1961)*708-728]

922. Residual spaces of TSs and homotopy type
δ/@   [PLMS 11,691-707]

923. Tauberian theorems for Dirichlet series over BSs
δ/A/@ [PLMS: 3(1953)378-384;(2)54,94-110(LT);Duke MJ 14,483-502]

924. Local homotopy invariants
δ/A   [PLMS(3) 3,351-367]

925. Summability of Laplace-Stieltjes integrals
δ/A   [PLMS 3,267-305]

926. Implicit integral operators
δ/@   [PLMS 11,434-456]

927. Plane-covering domains and vibrating membranes
δ/A/A [PLMS 11,419-433]

928. Iteration of operations over Boolean algebras
δ/@ [PLMS 11,385-401]

929. Extensions of Holder's inequality
δ/A [PLMS 11,311-326]

930. (Co)homology theory over noncommutative structures
δ/@ [PLMS: 11,239-275;12,1-29]

931. Integration over noncommutative structures
δ/@ [TAMS 68(1950)76-104;ALSO: SEGAL,I.M.(various),etc.]

932. Pseudo saddle points (generalized games)
δ/@ [PLMS 18,158-168]

933. Factorization of entire functions
δ [PLMS 18,69-76;GROSS,F.(bk);etc.]

934. Coefficient estimates for lacunary power-series
δ/A [PLMS 18,36-68 (Parts I,II)]

935. Allied sets in linear TSs
δ [PLMS 18,653-690]

936. Generalizations of the Euler polyhedra identities
δ/@ [PLMS 18,597-606]

937. Bi-topological spaces (quasi-metrics)
δ/A/@ [PLMS 13,1-19]

938. Definition of LTSs VIA NETS (using indecomposability)
S/A/@ [PLMS 13,1-19]

939. Relation of {f(nh)} to the general behaviour of the fn f
S/A/@ [PLMS 13,593-604]

940. Solutions of abstract differential inequalities
S/A/@ [PLMS 14,74-82]

941. Expansions of generalized fns in series of orthogonal fns
S/A/@ [PLMS 14,45-52]

942. Iteration products of infinite matrix transformation
S/@ [PLMS 14,342-352]

943. Real analogues of the 'identity theorem for analytic fns'
S/A [PLMS 14,245-259; JLMS 38(1963)91-98]

944. Interpolation between invariant measures
S/A [PLMS 14,208-220]

945. Extn. of topologies from {free generators} to full algebra
S/A [PLMS 14,566-576]

946. Convolutions of linear functionals
S/@ [PLMS: 14,431-444;15,81-104]

947. Embedding theorems for semigroups
δ   [PLMS 12,511-534]

948. Embedding of polcyclic groups
δ   [PLMS 12,496-510]

949. Minimum models of (small) entire/subharmonic functions
δ/A/@ [PLMS 12,445-495]

950. Representation of ?Hilbert-space integral oprators
δ/@   [PLMS 12,385-399]

951. (AP) closures of sequences of trigonometrical polynomials
δ/A   [PLMS 12,690-706]

952. Enumeration of triangulations of the disc
δ   [PLMS 14,746-768]

953. Regular neighbourhoods in TSs/manifolds
δ/A   [PLMS 12,719-745]

954. n- dimensional approximations to (classes of) functions
A   [PLMS 14,577-594]

955. Iteration formulae for fractional differences
δ/A/@ [PLMS 13,430-460]

956. Topologies on Boolean algebras
δ/A/@ [PLMS 13,413-429]

957. Generalizations of Chebychev's inequality (prob.theory)
δ/A/@ [PLMS 13,385-412]

958. Generalizations of 'positive-definiteness' for functions
δ/A/@ [PLMS 15,373-384]

959. Generalizations of 'second-order stationarity'
δ/A/@ [PLMS 15,385-398]

960. Integration over vector-valued measures
δ [PLMS 15,193-225]

961. Iterated integration over paths in R^n
δ/@ [PLMS 4,502-512]

962. Sublinear functionals in partially ordered (T)VSs
δ/A/@ [PLMS 4,402-418]

963. Transform theory and Newtonian interpolation
δ/A [PLMS 4,385-401]

964. Determinantal criteria for meromorphic functions
δ [PLMS 4,257-302]

965. Geometrical properties of fractional-dimensional plane sets
δ/@ [PLMS 4,257-302]

966. Relative densities for sequences of integers,etc.
δ/A [HALBERSTAM/ROTH, Sequences (OUP)* ;MANN,H.B.(bk)*]

967. Homotopy extension theorems
   S/A   [PLMS 6,100-116]

968. Proximity relations for ideals over plane curves
   S/A/@ [PLMS 6,70-99]

969. 'Laurent factorization' of operator-valued functions
   S/@   [PLMS 6,59-69;J.Indian M S 16(1952)25-30]

970. Rational approximation of operator-valued functions
   S/A/@ [PLMS 6,43-58]

971. p-length of p-solvable groups
   S/A   [PLMS 6,1-42]

972. Properties of transfinite dimension
   S/@   [PLMS 4,177-195]

973. Embedded dynamical systems
   S/A/@ [PLMS 4,168-176]

974. Embedding of algebraic systems
   S/A   [PLMS 4,138-153]

975. Abstract analogues of moment problems
   S/A   [PLMS 4,107-128]

976. Linear functions on domains of countably-inf. dimension
   S/A   [PLMS: 5,238-256;6,480-500]

977. Construction of integrals over parametric surfaces
δ   [PLMS 7,616-640]

978. Extemal problems related to systems of ODEs
δ/A/@ [PLMS 7,568-583]

979. Analogues of Taylor's theorem for algebraic functions
δ/A/A [PLMS 7,549-567]

980. Euclid's algorithm in algebraic function fields
δ/@ [PLMS 7,498-509]

981. Cesaro-Perron AP-function scale
δ/A/@ [PLMS 7,481-497]

982. Associative operations on groups
δ   [PLMS 6,581-596]

983. Representations of positive-definite functions
δ/A [PLMS 6,548-560]

984. Fourier-series problems req. very general forms of INTEGRAL
δ/A/@ [PLMS 1,46-57; (Bk;CUP c1988(?KONER)]

985. CV-analogue of Hardy-Littlewood relns.for f,f',f''
δ/A/A [PLMS 1,28-45]

986. Embedding of nonassociative rings in division rings
S/A [PLMS 1,241-256]

987. Some generalizations of matrix commutativity
S/@ [PLMS 1,222-231]

988. Almost-ordered groups
S/A/@ [PLMS 1,188-199]

989. Extension of Fredholm theory to functional equations in BSs
S/@ [PLMS: 53,109-124; 1,327-384]

990. Absolute Cesaro-summability of integrals
S/A [PLMS 1,308-326]

991. Well-ordered subseries of general series
S/A/@ [PLMS 1,291-307]

992. 'Locally','countably',and,'almost' free groups
S/A/@ [PLMS 1,285-290]

993. Generalized nilpotent transformations
S/@ [PLMS 1,494-512]

994. Min{ $Q(x,y)Q(u,v):|xv-uy| = 1$ }
S/A [PLMS: 1,257-283(Parts I,II,III);1,385-414,415-434]

995. Construction of irreducible invariant matrices
S [PLMS 2,98-127]

996. First-category-small sets of entire functions
δ/A [PLMS 2,60-68]

997. Geometric models of m-valued logics
δ/A [PLMS 2,30-44]

998. Surface charge distributions near edges
δ/A/A [PLMS 2,440-454]

999. Special fibrings of Riemannian manifolds
δ [PLMS 3,1-19]

1000. Mappings which do not increase small distances
δ/A [PLMS 2,272-278]

1001. Complex singularities of solns of generalized Lienard ODEs
δ/@ [PLMS 3,498-512]

1002. Contractions and expansions of n-ary operations
δ/@ [PLMS 8,127-134]

1003. Near-(semi)rings
δ/A/@ [PLMS: 8,76-94,95-108]

1004. Decomposition of linear oprs in completely positive cones
δ [PLMS 8,53-75]

1005. Topological-group-valued outer measures
δ/@ [PLMS 19,81-106]

1006. Independence spaces
δ   [PLMS 19,17-30]

1007. Dual orderings of BSs
δ/A   [PLMS 19,269-288]

1008. Convolution-smoothed functions
δ   [v Nostrand M Stud.#24 (SHAPIRO,H.S.)]

1009. Kobyashi pseudo-distance
δ/@   [MR 54 #604]

1010. Systems of equations over finite fields
δ   [LNM #536 (SCHMIDT,W.M.)]

1011. Probability theory on Boolean algebras of events
δ/@   [ONICESCU/CUCULESCU (bk;Inst.);PATHASARATHY (bk)*]

1012. Probability theory based on Radon measures
δ/A/@ [TJUR,T. (bk;?Springer)]

1013. Probability measures on l.c.groups
δ/A/@ [HEYER,H.,Ergebn.#94(Inst.);etc.]

1014. Diffusion processes and Riemannian geometry
δ/@/A [Usp.Mat.Nauk 30(1975)3-59;MR 54,#1404]

1015. Banach lattices and positive operators
δ/A/@ [SCHAEFER,H.H., bk?*(Grundl. Bd 215(1974);MR 54 #11023+]

1016. Theory of uniform algebras
δ [STOUT,E.L.(Bogden/Quigley,1971;MR 54 #10066;GAMELIN,bk]

1017. Functional 'central limit theorem' in BS
δ/A/@ [Ann.Prob.4(1976)600-611; MR 54 #11427]

1018. Computable ODEs with NONcomputable solutions
δ [J.Symbolic Logic (Reviews) ...]

1019. Algorithms for embedding an order into a maximal order
δ [LNM #353,pp204-221;MR 53 #358]

1020. Almost-positive matrices
δ/A [DONOGHUE,Grundl.#207,Ch.15]

1021. Block-Gerschgorin theorem (eigenvalue bounds)
δ/A/@ [Lin.Alg.Appl.13(1976)45-52 MR 53 #480]

1022. Abstract formulations of 'separation of variables'
δ/@ [Proc.R S Edin.A72(1974)57-78; MR 53 #1310]

1023. A generalization of 'metric space'
δ/A/@ [Soviet Math.(Iz.VUZ)19(1975)(4)26-28;MR 53 #1553]

1024. Generalizations of the Weierstrass approximation theorem
A/@ [Bull.Soc.Mat.France Suppl.Mem.#38(1974)]

1044. Factorization of meromorphic functions
    δ   [GROSS,F.(Naval Res.Lab.Rep.,1972);MR 53 #11030]

1045. Stochastic analogues of fixed-point theorems
    δ/A/A/@ [BAMS 82(1976)641-657; MR 54 #1390]

1046. Variational approximation of eigenvalues
    δ/A   [WEINBERGER,H.F.(bk;SIAM,1974);MR 53 #3842]

1047. Restrictions of analytic functions
    δ   [PAMS: 48,113-119;51,335-343;52,222-226;MR 53 #3764a-c]

1048. Distance geometry (metric topology)
    δ/A/@ [BLUMENTHAL,L.M.(OUP,1953)*]

1049. The role of nearness spaces in topology
    δ/A   [LNM #540,1-22 (Inst.)]

1050. Nearness -- in topology, and in general mathematics
    δ/A   [LNPAM #24 (Dekker)77-86]

1051. Bounded,almost-bounded,and,nearly-bounded,sets
    δ/A/@ [MR 56 #13167 ...]

1052. Algebraic calculus (nilpotent elts.adjoined to R)
    δ/A   [KOCK,A.,LMS Lec.Notes #51(CUP)*]

1053. Formulae for the n-th derivative of composite function
    δ   [Math.Gaz.: 54(1970)386;56(1972)39;PCPS (GOOD,I.J.)]

1054. Fractional differentiation
δ/@ [LNM #457; MR 5811267; ?MR 51 #3369]

1055. Justification of 'averaging' for (nonlinear) (O)DEs
δ/A [PLMS 34(1977)385-420; MR 58 #22829]

1056. Construction of Volterra kernels(noncomm. oprnl calculus)
δ/@ [FLEISS/LAMNABI (LNCS/SAM)]

✶ ALL OF ## 1057 - 1263 REFER TO J.CHEM.PHYS. (JCP)

1057. Ising models on compressible lattices
δ/A [JCP 44,1120-1129]

1058. Metrics/inequalities for 4-particle configurations *1108*
δ/A/A [JCP 49,3683-3687]

1059. Diffusion-controlled chemical reactions *432*
δ/A [JCP 49:701;3618-3627]

1060. General deformation of molecules *1272*
δ/A/A [JCP 49,1510-1520]

1061. Combinatorial aspects of melting ... of DNA *1284*
δ/A [JCP 44,4567-4581]

1062. Mathematical formulation of resonance conditions *1279*
δ/A [JCP 44,4431-4439]

✶ Extracted from the full list {1, ..., 1398} [conversion]

1063. Flow of oriented fluids   *1290*
δ/A   [JCP 56,5063-5071]

1064. Master equations for quenched linear chains   *1287*
δ/A   [JCP 56,4852-4867]

1065. Phase transitions for 'hard ellipses'   *1285*
δ   [JCP 56,4729-4744]

1066. Unimolecular interactions (survey)
δ/A   [JCP:36,1466- ;38,1959-1966]

1067. Consistency criteria for truncation of BBGKY hierarchy   *770*
δ/A   [JCP 50,5334-5336]

1068. Calculation of 3rd virial coeff. for nonspherical molecules   *782*
δ/@   [JCP 50,4967-4986 (?SAM)]

1069. Collision integrals (general scheme ...)   *779*
δ/A/@ [JCP 50,4823-4831 ;H,C&Bird(bk)*]

1070. Random walks and nucleation   *768a*
δ/A   [JCP 50,4672-4678]

1071. Light-scattering from oscillating chemical systems   *1111*
δ/A   [JCP 57,4327-4332]

1072. Chemical reactions in imperfect gases   *1164*
δ   [JCP 51,252-255]

1073. Mathematical models for photochemical dissociation *1152*
   δ   [JCP 60,2370-2383]

1074. Mathematical(stability/...)properties of reacting systems *216*
   δ/@ [JCP 59,4164-4173]

1075. Interacting molecular chains
   δ   [JCP 59,4029-4034]

1076. Approximate n-dimensional numerical integration *1124*
   δ/A/@ [JCP 59,3992-4008]

1077. Self-avoiding random walks *219*
   δ   [JCP 59,5787-5795]

1078. Generalized molecular distribution functions *201*
   δ/@ [JCP 59,6535-6555]

1079. Formal calculations for anharmonic oscillators *154*
   δ   [JCP 59,6433-6449 (?SAM)]

1080. Phase transitions(T-D limits)for bi-polymers *728*
   δ   [JCP 59,5605-5614]

1081. Diffusion-controlled processes
   δ/A [JCP:40,3040;47,3049-3061]

1082. Non-additive interactions *86*
   δ/A/A [JCP 47,2889-2897]

1083. Energy-moment methods ... *81*
    δ   [JCP 47,2784-2792]

1084. Clusters, surface tension, and critical phenomena *102*
    δ/A   [JCP 47,2513-2533]

1085. Asymptotics for QM anharmonic oscillators *1091*
    δ/A   [JCP 50,1182-1194]

1086. Helix-coil phase transitions *5*
    δ   [JCP 48,5223-5230]

1087. Anisotropic interfaces *604*
    δ   [JCP:51,4983- ;51,5363-5373]

1088. Statistical T-D of chemical reactions *401*
    δ/A   [JCP 51,5193-5203]

1089. Theory of reactive scattering *606*
    δ/A   [JCP 51,5204-5215]

1090. Long-range multi-molecular interactions *558*
    δ/@   [JCP 56,1238-1246]

1091. Pade approximations for the harmonic oscillator *157*
    δ   [JCP 56,11041112]

1092. Random-walk models of adsorption 555
S/A   [JCP 44,2130-2138]

1093. Kinetics of unimolecular breakdown 550
S/A   [JCP 44,2029-2053]

1094. Even moments(?of position) for polymer chains 178
S   [JCP 48,5646-5655]

1095. Rotational Brownian motion ... and DNA
S/A   [JCP 51,4790-4793]

1096. Mathematical resolution of 'paradoxes' for nucleation 544
S/A   [JCP 48,5553-5560]

1097. Surface properties of liquids(mixtures) 540
S   [JCP 48,5361-70]

1098. Interface between immiscible polymers 234
S   [JCP 56,3592-3601]

1099. Thermodynamics with internal state variables 844
S/@   [JCP 47,597-613]

1100. Thermodynamics of nonuniform solutions 807
S   [JCP 56,4043-4053]

1101. Exact solution of coupled (Hartree-Fock) equations 802
S   [JCP 56,3823-3831]

1102. Correlated-molecular-rotation models 608
δ   [JCP 51,5070-5089]

1103. Sequences of first-order reactions 607a
δ/@/A   [JCP 51,5028-5033]

1104. Numerical solution of coupled integral equations 598
δ/A   [JCP 51,4809-4819]

1105. Formal mathematical schemes for polyatomic fluids
δ/A/A   [JCP:42,3445- ;43,1750- ;45,3485-3513]

1106. EXCITONS in liquids
δ/A   [JCP:44,4470- ;46,4445-4453] 479a

1107. Expansions(Special fns./series)of various integrals 474
δ/A   [JCP 46 4362-4380]

1108. Graph-theoretic expansions of virial coefficients 479
δ/A   [JCP 46,4181-4197 (?SAM)]

1109. 'Surface tension' of a lattice gas 484
δ   [JCP 57,847-851]

1110. Boundary free energy in lattice models 163
δ   [JCP 57,777-791]

1111. Two-phase fluid interfaces
δ   [JCP 57,769-776]

1112. Fugacity expansions for ... adsorption 470
δ/A   [JCP 57,667-675]

1113. Diagrammatic formalisms(?general) 471
δ/@/A   [JCP 57,638-666]

1114. Stochastic models for chemical reactions 231
δ/A   [JCP 45,2145-2155]

1115. Kinetics of helix-coil transitions 166
δ   [JCP 45,2071-2090]

1116. Kinetic theory of dense fluid mixtures 489
δ   [JCP 45,2020-2031]

1117. Interaction potentials(for periodic linear arrays) 488
δ   [JCP 45, 1941-1945]

1118. Irreducible cluster integrals for hard spheres 486
δ   [JCP 45, 1866-1874]

1119. Mayer theory of condensation/nucleation 816/817
δ   [JCP 48,732-740; 48,776-780]

1120. Trancriptions for decorated-lattice models 815
δ/@   [JCP 48,741-748]

1121. Deterministic vs. stochastic models of chemical reactions
δ/A   [JCP 57,2976-2978(and refs)]

1122. Statistical mechanics of 'worm-like chains'   507
δ   [JCP 57,2839-2854]

1123. Asymptotics for Green/Kubo integrands ...   509
δ   [JCP 57,2800-2810]

1124. Bifurcation-co-ordinates for chemical reactions   172
δ/@/A [JCP 57,2722-2727]

1125. Montecarlo calculations for hard discs   520
δ   [JCP 48,415-434]

1126. Viscoelasticity of networks   1424/523
δ/A   [JCP:45,1505- ;48,235-243]

1127. 'Brownian carriers' ...   527
δ/A   [JCP 48,120-127]

1128. Quantum mechanical equations for chemical reactions   530
δ/@   [JCP 48,56-74]

1129. Effective interaction potentials for QM ideal gas
δ   [JCP 47,5369-5375]

1130. QM transport coefficients   329
δ/@   [JCP 47,5269-5289]

1131. Theory of the helix-coil transition 326
  δ   [JCP 47,5116-5128]

1132. Kinteic theory of polyatomic gases 322
  δ   [JCP 47,4978-4993]

1133. Stress tensors(?various models) 321
  δ   [JCP 47,4959-4971]

1134. Deviations form pairwise-additivity 319
  δ/A [JCP 47,4916-4921]

1135. Statictical mechanics of hard-particle systems 568
  δ   [JCP 48,3139-3155]

1136. Adsorption of flexible macromolecules
  δ/A [JCP:48,4835-4851;36,1872- ;42,2101- ;43,2392- ;46,1105-] 706

1137. Unimolecular breakdown ... 582
  δ/A [JCP 44,3597-3607]

1138. General mobility theory(for solutions) 295
  δ   [JCP 57,3749-3762]

1139. Thermodynamics of curved boundary-layers 294
  δ/A [JCP 57,3745-3748]

1140. 'Mixture models' for classical fluids 349
  δ   [JCP:56,2864- ;57,3605-3612]

1141. Theory of reactive scattering 261
S/A   [JCP 57,3441-3455]

1142. 'Scaled-particle theories' for rigid discs 305
S   [JCP 57,3556-3578]

1143. Optical/accoustic responses in multi-level systems 301
S   [JCP 57,3130-3145]

1144. Evaluation of various integrals 299
S   [JCP 57,3029-3044]]

1145. Reduced density matrices
S   [JCP:46,2752- ;47,2298-2311] 266

1146. Nonexistence of perimetric co-ords. for (>4) particles
S/A   [JCP 47,2229-2231]

1147. Computations with pseudo-potentials 252
S/@   [JCP:42,2363- ;45,2898-2912]

1148. Brownian motion of polyatomic molecules 248
S/A   [JCP 44,3988-4004]

1149. Integral equations for classical distribution functions 246
S   [JCP 44,3941-3946]

1150. Constructive operator solns of time-dept. Schrodinger eqs. *277*
S [JCP 44,3897-3912]

1151. Random packing of hard spheres *275*
S/A [JCP 44,3888-3894]

1152. Bounds on partition fns. for finite hard-particle systems
S/A [JCP 44,3572-3578]

1153. Stochastic models of chromotography *394*
S/A [JCP 51,3886-3890 (Oxtoby!)]

1154. Inverse problems for potential scattering (WKB) *158*
S/A [JCP 51,3631-3638]

1155. Analytical evaluation of various integrals
S [JCP:51,4287- ;48:4098- ,4106- ,4108- ;51,956-]
        160   *932*   *637*

1156. Brownian motion and surface chemical reactions *1209*
S/A [JCP 60,1087-1093]

1157. 'Natural iteration' in cluster-variation schemes *1208*
S/A [JCP 60,1071-1080]

1158. Macromolecular interactions and diffusion *1205a*
S/A [JCP 60,976-989]

1159. Diffusion-controlled intra-chain reactions of polymers *1200*
S/A [JCP 60,866-890]

1160. Linear-chain partition functions *1240*
   δ   [JCP 60,808-812]

1161. Exact solution of coupled differential equations *1239*
   δ   [JCP 60,760-765]

1162. Graph-theoretic stability analysis of reaction networks
 δ/A/A [JCP:1481-1492;60,1493-1501]
         60 *1196*

1163. Functional variation of partition functions *1187*
   δ   [JCP 60,1197-1207]

1164. Phase transitions for 'long hard rods' *536*
   δ   [JCP 59,2596-2601]

1165. Insensitivity(to wave-fns)of psuedo-potl.calculations *171*
  δ/A [JCP 49,679-691]

1166. Interconversion of chemical species *170*
  δ/A [JCP 60,3474-3482]

1167. Hyperspherical co-ordinates in atomic descriptions *174*
   δ   [JCP 60,66-80]

1168. Random-coil chains near obstacles *92*
  δ/A [JCP 59:3701- ;59,3799-]
         92

1169. Calculations for 'brick-wall arrays' of hard squares *378*
   δ   [JCP 59,3654-3658]

1170. Forces between ellipsoidal particles *38*
   δ   [JCP 58,2763-2766]

1171. Mathematical problems for expansions in lattice dynamics *37*
   δ   [JCP 58,2724-2745]

1172. Variational principles for 'excluded-volume problems' *185*
   δ/A   [JCP 47,186-194]

1173. Helix-coil transitions for 'block-copolymers' *24*
   δ/@   [JCP 45,917-927]

1174. Polymer dynamics(general)
   δ   [JCP:45,785-803;44,2107- ;42:3831- ;3838-] *83/553*

1175. Symbolic expressions for pressure-broadening(?spectrum) *651*
   δ   [JCP 51,2192-2205]

1176. Fine-structure studies/optical levels(formulae) *652*
   δ   [JCP 51,2152-2170]

1177. Theory of 'absolute reaction rates' *180*
   δ   [JCP 47,1235-1247]

1178. Probabilistic treatment of polymers *721*
   δ   [JCP 58,1553-1568]

1179. Dynamics of polymers in dilute solutions 720
   δ   [JCP 58,1459-1466]

1180. Tensors invariant under crystal point groups 708
   δ/@ [JCP 58,468-471]

1181. Helix-coil transitions for periodic copolymers
   δ/A [JCP 46,1037-1042]

1182. Models for calculation of chemical reaction cross-sections
   δ/A [JCP:45,2630- ;46,959-966]
          924   198

1183. Relaxation of a 'string oscillator'
   δ   [JCP 50,4324-4335]

1184. Bulk properties of two-phase random media
   δ/@ [JCP50,4305-4312]

1185. Structural calculations for chain polymers 682
   δ   [JCP 50,4165-4200]

1186. Mulipole inequalities ... 680
   δ   [JCP 50,4136-4137]

1187. Fluctuations in open systems 741
   δ   [JCP 51,2632-2637]

1188. Field-theoretic methods in stat. mech. 205
   δ/A [JCP 51,2410-2418]

1189. Green's functions in molecular calculations 892
   δ   [JCP 49,2002-2025]

1190. Nonequilibrium unimolecular reaction theory 80
   δ/A [JCP 45,4280-4288]

1191. Phase transitions for hard squares (nearest-nbr. exclusion) 134
   δ   [JCP 45,3983-4003]

1192. Dynamics of dissociation ... 343
   δ/A [JCP 56,2582-2591]

1193. Viscous flow past randomly-placed spheres 342
   δ/A [JCP 56,2527-2539]

1194. Statistical mechanics of 'simple entanglement' 373
   δ   [JCP 46,1475-1483]

1195. Irreversibility in finite quantum-mechanical systems 95/94
   δ/A [JCP:46,1365-1372;45,1352- ;Amer.J.Phys.:33,722-;34,411-]
              95   94   ?   ?

1196. Isolated polymer molecule at an interface 12
   δ   [JCP:43,1591- ;44,4264-4272]
                    12

1197. Genrlzn of Hamilton's Principle for cts dissipative 7
   systems
   δ/@ [JCP 59,2929-2936]

1198. Molecular viscosity in polymer dynamics
   δ

125

1198. [JCP 59,2858-2868] 3
δ

1199. Phase-space treatment of 'bound states' 31
δ/@/A [JCP 57,5577-5594]

1200. Reaction-lifetimes;collision-lifetimes 9
δ [JCP: 57,5562-5576;57,5418-5426]

1201. Rigorous dynamics of semi-infinite chains 13
δ [JCP 57,5037-5044]

1202. Asymptotic analysis for the QM anharmonic oscillator 130
δ/A [JCP 47,4540-4547]

1203. Rigorous determination of potential-energy surfaces 77
δ/@ [JCP 59,1959-1973]

1204. Brownian motion of N interacting particles 337
δ [JCP:57,2098- ;59,1833-1840]

1205. Finite-diff. methods for power-series density expansions
δ/A [JCP 57,2700-2709]

1206. Cluster expansions (and reductions) for fluids ... 330
δ [JCP 57,1918-1937]

1207. Correlation functions for 1-D sq-well gas (Laplace trans.) 936
δ [JCP 48,4246-4251]

1208. Liapounov stability for chemical reactions (simplex)
S/A [JCP 48,4144-4147]

1209. Analytical evaluation of various integrals 932
S [JCP 48,4098-4116 (?SAM)]

1210. Independent relaxation in multicomponent fluids 920
S [JCP 45,2492-2507]

1211. Exact matrix calculational scheme for lattice gases 919/959
S [JCP:45,2482-2492;47,4015-]

1212. SHAPE of self-avoiding random walks (polymers) 915
S/A [JCP 44,616-622]

1213. Recombination-dissociation kinetics ... 913
S/A [JCP 44,582-595]

1214. Stochastic analysis of multicomponent chemical reactions 904
S/A [JCP 44,990-997]

1215. Stochastic and deterministic reaction-rate equations 1023
S/A [JCP 50,460-466]

1216. Evaluation of collision integrals for hard spheres 944
S [JCP 56,5583-5601]

1217. Enumeration of permutations of isomerization reactions
S

1217. [JCP 56,5478-5489] 942
δ

1218. Polymer dynamics ...
δ   [JCP 44,2595-2602] 1035

1219. Excluded vol.in linear polymer chains(hierarchy of DEs) 958
δ   [JCP 47,3991-3999]

1220. Approximation of partition functions for crystals 949
δ   [JCP 47,3763-3771]

1221. Thermochemical reactions ... 1274
δ   [JCP 49,1625-1637]

1222. Quenched linear chains 1287
δ/A   [JCP56,4852-4867]

1223. Hard-ellipse systems 1285
δ   [JCP 56,4729-4744]

1224. Kinetics of growth of multicomponent chains 1301
δ   [JCP 47,3451-3469]

1225. Reversible isomerization/termolecular reactions 1294
δ   [JCP 47,3276-3296]

1226. Boundary perturbations in unstable chemical reactions 1310
δ   [JCP 56,287-294]

1227. 'Classical quantization' of nonseparable systems
δ/A   [JCP 56,38-45]

1228. Construction of effective pair-potentials 973
δ   [JCP 57,1780-1787]

1229. Effective potentials between linear molecules 970
δ   [JCP 57,1718-1725]

1230. Automatic generation of molecular symmetry co-ordinates 968
δ   [JCP 57,1616-1620]

1231. Diffusion of plane polymers 966
δ   [JCP 57,1537-1546]

1232. Droplet growth in nucleation ... 965
δ   [JCP 57,1441-1458 (?SAM)]

1233. Viscoelasticity of polymers 1424
δ/A   [JCP 45,1505-1514]

1234. Unimolecular rate theory
δ/A   [JCP:45,1444-1451;32,677- ; bks.:NIKITIN(OUP)*,SLATER*]

1235. Decay of initial correlations in stochastic processes 1086/
δ/A   [JCP:46,4100-4111;50,3663- 1/2   1059
        1086   1059

1236. Analytical estimates of excluded volume 1082
δ   [JCP46,3789-3810]

1237. Functional-integral reprns.in nonequilibrium stat.mech. *193*
S   [JCP 46,3707-3717]

1238. Dependence of time-correlation fns.on (pair-)potentials *1074*
S   [JCP 48,2494-2501]

1239. Bounds for solutions of the 1-D Poisson equation *1067*
S/A   [JCP 50,3921-3924]

1240. Random-network models for rubber ... *1066*
S/A   [JCP 50,3889-3903]

1241. Decay of correlations in linear systems *1062*
S   [JCP 50,3756-3772;Phys.Rev.156(1967)583(Sec.4.2)]

1242. Rigorous cluster distributions for 1-D lattices *1060*
S   [JCP 50,3707-3717]

1243. Random-matrix master equations *1049*
S   [JCP 57,4699-4712]

1244. Various BOUNDS in the theory of classical fluids *1370*
S/A   [JCP 58,2223-2229]

1245. Quasi-statistical complexes in chemical reactions *1385*
S/A   [JCP 56,5786-5796]

1246. SPAN of a polymer chain *1382*
δ/A [JCP 56,5747-5757]

1247. Classical/quantum mechanics of linear collisions *1328/1333*
δ [JCP:45,4493-4504;41:603,610,2614,2624;43,1598-;49,2610-]
   1328 ?   ? *1333*

1248. Growth of multicomponent(chemical)chains *1325*
δ/A [JCP 45,4444-4454]

1249. Exponents for Gaussian atomic orbitals *1323*
δ [JCP 45,4400-4413]

1250. Approximation techniques for(lattice)gases *1041*
δ/A [JCP 50,2247-2254]

1251. Recursive determination of cluster expansions *990*
δ [JCP 50,1928-1934]

1252. Hard spheres with surface adhesion *1004*
δ [JCP 49,2770-2774]

1253. Steepest-descent calculations of 'complexions'(St.Mech.) *1418a*
δ/A [JCP 49,4187-4192]

1254. Analysis of QM anharmonic oscillators *1358*
δ [JCP 50,3342-3354;50,1182-1194]

1255. Macroscopic quantum electrodynamics *1359*
δ/A [JCP50,3355-3377]

1256. Curvature/torsion-dependent polymer configurational energy
δ/A [JCP 50,3137-3142]   *1353*

1257. Thermal unimolecular reactions *1342*
δ   [JCP 45,216-223]

1258. Use of inequalities to evaluate correlation functions *177*
δ/A [JCP 55,5101-5109]

1259. Calculation of one-center Slater orbitals(integrals) *1251/*
δ   [JCP 55,4699-4710;51,3434-3447]   *1270*
         1251

1260. Exact partition fn.for polymer chain-folding(crystalzn) *457*
δ   [JCP 48,3351-3360]

1261. Graph-counting and nonadditivity in statl.mechanics *448*
δ/A [JCP 49,72-80]

1262. Irreversibility and information *94*
δ/A [JCP 45,1352-1357]

1263. Rigorous relaxation calculations *1262*
δ/A/A [JCP 45,1105-1112]

1264. Constructive counter-example in Q-M scattering theory
δ   [PEARSON,D.B.(prepr.);later,?Comm.Math.Phys.   ]

1265. Abstraction/idealization/approximation in phys.theories
    δ/@/A   [Proc.,HARTKAMPER,A.,ed. (Plenum P,1981;pp264)*-]

1266. {Recursive languages} as a TS (complexity theory)
    δ/A   [REGAN,K.(prepr.(Merton,c1986)); ?bks ...]

1267. Mean-value theorems in recursive function theory
    δ/A/A   [PLMS 52,81-107*]

1268. 'Discrete convergence' of continuous maps in MSs
    δ/A/@   [STUMMEL,F.,various preprints*]

1269. Constructive analysis/approximation theory
    δ/A   [BRIDGES,D.S.,(p)reprints*;ALSO,by other authors*]

1270. Iterative parallel solution of tri-diagonal eq.systems
    δ   [JACM 23,636- (HELLER,et al.)*;TRAUB (Conf.Proc.)*]

1271. Mathematical problems related to Hilbert's '23 Probs.'
    δ   [AMS Proc.Sympos.P M #28 *--]

1272. Constructive solution of Hilbert's 17th Problem
    δ/@   [DALZELL ,Thesis (Stanford)*]

1273. Applications of differential forms
    δ   [FLANDERS,H. (bk)*]

1274. Topological proofs of function-theoretic results
δ/@/A [WHYBURN,G.,Topological Analysis(?Princeton)*-]

1275. Applications of orthogonal families of functions
δ [EPSTEIN (bk)*-]

1276. Applications of Nonstandard Analysis
δ/@ [DAVIS,M.(?McG-H)*-]

1277. Evaluation of various Wiener integrals
δ/A [Papers:CAMERON/MARTIN*;FOSDICK* ...]

1278. Point-sets of harmonic measure zero
δ/A [NEVANLINNA,R.(Springer),part*.]

1279. Various generalizations of 'topological space'
δ/A/@ [Gen.Topol.Appl.:5,191- ;263- ;8,111- ;9,233- ;1,65-;2,1]

1280. Development of set-theoretical topology(Survey)
δ/@/A [Russian Math.Surveys 33,1-53*]

1281. Generalized f-contractions
A/@ [Mathematica(?USSR)24(47)1/2(1982)175-178*]

1282. Convex sets in spaces of analytic functions
A/δ [Mathematica 24 ... ,111-116*]

1283. Metrical fixed-point theorems
A/@ [Mathematica 24 ... ,85-98*]

1284. Optimality criteria for polygonal convex programming
   A/A   [Mathematica 24 ... ,65-68*]

1285. Probabilistic pseudometrics
   A/@   [Mathematica 24 ... ,21-29*]

1286. Mildly nonlinear variational inequalities
   A/@   [Mathematica 24 ... ,99-110*]

1287. Generalized kernels of digraphs
   @/A   [Mathematica 24 ... ,57-63*]

1288. 'Order of starlikeness' for(analytic)functions
   δ   [Mathematica 24 ... ,73-78*]

1289. Properties of fuzzy topological spaces
   A/@   [J Math Anal Appl:56,621-633*;58,559-571* ;
          Infn.Sciences 14,107-113*;Genl.Topol Appl.:
          10,147-160* ;11,59-67*;Fuzzy Sets & Systems
          3,93-104;PLUS:various (p)reprints*]

1290. Controllability/stabilizability for linear PDEs
   δ/A   [SIAM Rev.20(1978)639-739* (RUSSELL,D.L.)]

1291. Use of fp-theorems in global nonlinear control theory
   A/@   [Warwick U.Reps./preprints ##87,89,98(PRITCHARD,A.J.)

1292. Generalization of the(?Popov) Circle Criterion(Control Th)
   A/@   [Oxford Engrng,Sci.Rep.(KOUVERATAKIS,B.)*]

1293. Use of 'Walsh functions' in analysis of limit cycles
δ/A   [Int.J.Control(IJC) 32,663-681*]

1294. Stabilization of linear oscillators in HS
δ/A/@ [J.Math.Anal.Appl.25(1969)663-675*]

1295. Nonuniform decay rates for HS oscillations
δ/A/@ [J Diffl.Eqs(JDE) 19,344-370*]

1296. Random Harmonic Analysis
δ/A/@ [ROSENBLATT,M.(bk)*-]

1297. Statistical analysis of stationary processes
δ   [ROSENBLATT,M.,Conf.Proc.*]

1298. Nonlinear Prediction(Survey)
δ/A   [MASANI.P./WIENER,N.,Conf.Proc.*]

1299. Integral reprns.of transition-probability matrices
δ/@   [KENDALL,D.G.,Conf.Proc.*]

1300. Nonlinear problems in probability theory
δ/@   [GRENANDER,U.,Conf.Proc.*]

1301. Monotone matrix functions and Wold decompositions
δ/@   [MASANI,P.,Conf.Proc.*]

1302. Deterministic/stochastic approximation of regions
δ/A/@ [DAVIS,P.J.,et al.,J.Approx.Th.21,60- *]

1303. Stochastic Processes and Filtering Theory
δ/@   [JAZWINSKY (bk)*]

1304. Simulation-relations for dynamical systems
δ/A   [?KULIK,Rep.Waterloo CS-80-6*]

1305. Factorization 'in polynomial time'(algorithms)
δ   [COLLINS (SAC-1),Conf.Proc.*]

1306. Solution of classes of integral equations (REDUCE)
δ   [LOOS,R.(U.Utah Prepr.)*]

1307. Simultaneous approx.of PS-solutions of alg.fnl.equations
δ/A/@ [OSGOOD ,Amer.J.Math.103(1981)469-477*]

1308. 'Liouvillian solutions' of n-th-order ODEs(constructive)
δ   [Amer.J.Math.,103,661-682*]

1309. Algebraic('field')properties of Elementary functions
δ   [Amer.J.Math.101,743-759*]

1310. Use of integral transformations to solve 'F(x) = 0'
δ/A   [Traub,ed.,Conf.Proc.(KACEWICZ)*]

1311.'Integration in finite terms' for transcendental integrands
δ   [TAMS 139,167-189 (RISCH,R.)*]

1312. Aspects of Symbolic Integration and Simplification
δ   [ROTHSTEIN,Thesis (U.Wisconsin,1976)*]

1313. Simplification of nested radicals
  δ   [SHTOKHAMER,R.(U.Utah,1975,Prepr.)*]

1314. 'Implicitly elementary integrals are explicitly elementary'
  δ   [PAMS 57,1-   (RISCH,R.)*]

1315. Purely ALGEBRAIC proofs of 'finite terms' integration thms
  δ/@ [Pacific J.Math.24,153-   (?SINGER)*]

1316. 'Canonical forms' of solutions of algebraic ODEs
  δ   [Pacific J.Math.59,535-   (SINGER)*]

1317. Number of solutions of equations in finite fields
  δ   [WEIL,A.,paper*]

1318. Schanuels Conjectures(transcl.properties of fns.)
  δ/A [AX,J.,paper*]

1319. Differentials(long survey;relations to BS-approximation)
  δ/A/@ [NASHED,M.Z.,paper*,in U.Wisconsin Conf.Proc.]

1320. Nonlinear optimization (constructive)
  δ/A [BELTRAMI,?E. (?bk)*]

1321. Qualitative methods for the n-body problem
  δ/A/@ [KHILMI bk (G&B)*]

1322. Approximation theorems in mathematical statistics
δ/A   [SERFTING,R.G.(Wiley,1981;pp371)]

1323. Baire-category criteria for 'usual', in rational approxn.
δ/A/@   [Canad.Math.Bull.26(1983)317-324]

1324. Iterative geometrical constructions
δ/@   [AMM 90,421-430?*]

1325. Inter-cell maps(decomposition of nonlin.dynamical systems)
δ/A   [HU,C.S.Int.J.Nonlin.Mech.(1983)?*; J.Appl.Mech(later)*]

1326. Inequalities VIA 'continuous dynamic programming'
δ/A/@   [J.Math.Anal Appl.:96,119-130,71,423-430;78,522-530;
         80,31-35;86,96-98]

1327. Reprns.of integers as product of integers of given types
δ   [J.Austral.M S A35(1983)143-162]

1328. Approximately continuous integrals (Burkhill)
δ/@   [J.Austral.M S A35,236-253]

1329. Composition of completely monotonic sequences/functions
δ   [JLMS 28,31-46]

1330. Inequalities for algebraic/trigonometrical polynomials
δ/A   [JLMS 28,83-93]

1331. Set-valued extensions of integral-ineqs.(interval anal.)
A/@   [j.Integral Eqs.5(1983)187-199]

1332. Fixed points in theory of computer algorithms/programs
δ/A/@ [e.g., Algebraic Semantics (CUP)*, and various papers*]

1333. Information theory(?/CT) and urban planning/development(?)
δ/A [WEBBER(?UBC), bk, Croom-Helm]

1334. Locally recurrent functions
δ/A/@ [AMM:70,822-826;72,983-985;84,191-195]

1335. Computation of ENTROPIES of various sets of functions
δ/@ [Canad.J M15(1963)422-432;Pacific J M 13(1963)1085-1095;
       MR 28 ##2190,2191]

1336. Constructive conformal mapping(n-connected domains)
δ/A [TAMS:82(1956)128-146;93(1959)81-96]

1337. Topological properties of quantized spaces(CS)
δ/A/A [JACM 18(4/71)239-255*]

1338. Inversion of the van der Monde matrix
δ [AMM 74,571-574]

1339. Finitely separable/approximable semigroups
A/@ [MR 28 #32151 (USSR)]

1340. Path-integrals in polar co-ordinates
δ/@ [Proc Roy Soc(PRS) A279(1964)229-235]

1341. SUMMATION in finite terms
  δ   [KARR,?M.;JACM(1981)*]

1342. Mathematical foundations of 'renormalization group'(SM)
 δ/A/A [DOBRUSHIN,LNP #80,303-311*;Phys.Reps.(BARBER,M.N.);long]

1343. Bounds on bi-harmonic fns.from boundary behaviour only
 δ/A   [Trans.24th Conf.Army Mathns.,1978ARO Rep.79.1,231-248]

1344. Approximation in the Hausdorff metric
 δ/A   [SENDOV,B.,bk(USSR);MR 80j #41004)]

1345. Approximation:general problems(BSs,splines, ...)
 δ/A/@ [HOLLAND/SAHNEY,bk (Krieger,1979;pp344;Biblio 1389 items]

1346. Distance from the truth(phil.of sci) (???)
 A/@!   [NINNUOLETTO,Truthlikeness (Reidel,?1988)&Biblio.]

1347. Asymptotically orhtonormal sequences in HS
  A   [MR 24A #2831]

1348. Kantorovich inequality
  A   [J Res NBS 64B(1960)33-34;MR 22A #45]

1349. The Volterra-Wiener theory of nonlinear systems
 δ/A   [SCHETZEN,M.,bk]

1350. Continuous analogues of Chebychev's inequality
 A/@/A [Th.Prob.Appl.(1958);Ann Mat.Stat.29,226-234;
          QJMath. 9,232-240;MR 22 #6619(multivariate)]

1351. { Manual of Mathematical Physics (Comprehensive summary)}
δ   [RICHARDS,P.I.(Pergamon P,1960)]

1352. Simultaneous (D)AC
δ/@   [JLMS 34,264-272*]

1353. NEUTRICES (generalized numbers) (asymptotic analysis)
δ/@   [J SIAM 7(1959)253-279 (?van de CORPUT)]

1354. Analogies between plane alg.curves & plane dynamic.systems
δ/A   [MR 22(1960) #7336]

1355. Scaled metric spaces ...
δ/A   [BAMS 79,566-569?*]

1356. MS of languages/topological 'learning space'
δ/A/A [LNCS #53,537-542*]

1357. 'Riemann surfaces'/integration of eqs.of motion
δ/@/A [GOLUBEV,V.,bk(Israel transl.,1953)*-;LEIMANIS(Springer)]

1358. Nec./suffic.conditions for fractional differentiability
δ/A/@ [PLMS 14A,249-264]

1359. Almost-finiteness properties(recursive functions)
δ/@/A [ROGERS,H,Jr.,Th.Rec.Fns ...(McG-H,1967;p240 ff]

1360. Reconstrunction of 2-D objs.form 1-D data(Radon trans.)
δ/A/@

[HELGASON,bk(Birkhauser,1980)]

1361. Uniform summability (various methods)
δ/@ [PCPS 71(1972)335-341*]

1362. MUTUALITY(generalization of duality)for NEARNESS-spaces
@/A [PCPS 71,203-210]

1363. Definitions of PRODUCT for elementary distributions(gfs)
δ [PCPS71,123-130;QJM 22(1971)291-298:?FISHER,B.]

1364. Stieltjes transforms of generalized fns.(gfs)
δ/@ [PCPS 71 85-96]

1365. Restricted approximation over classes of rational functions
δ/A [PCPS: 60,877-890;58,244-267]

1366. Negative- spectrum condns.for Fredholm oprs.in lc spaces
δ [PCPS 60,801-806]

1367. Simultaneous diophantine approximation
A/@ [PCPS 67,75-86]

1368. Functional Norlund summation ...
δ/A [PCPS 67,41-60]

1369. Summability of Fourier integrals
δ/A [PCPS 67,23-28]

1370. Distribution of sequences in abelian groups
δ/@ [PCPS 67,1-11]

1371. Time-structures for multivalued dynamical systems
δ/@ [PCPS 67,635-646]

1372. Random walks and electric current in networks
δ/A [PCPS 55,181-194]

1373. Generalized intervals in partially-ordered groups
δ/A/@ [PCPS 55,165-171]

1374. Decomposition of spaces in cartesian products/unions
δ [PCPS 55,248-256]

1375. Volumes of sets of matrices
δ/@ [PCPS 55,213-223]

1376. Algebraic (BS) construction of 3-D harmonic functions
δ/@ [PCPS 67,383-389]

1377. Determination of deformed paths for 'saddle-point-anal.'
δ/A [PCPS:67,?371-381;76,211-231;65,113-128;72,49-]

1378. Random Fourier-series transformations on P.O.groups
δ/A/@ [PCPS 67,295-306]

1379. Approximating-sequences/Hausdorff measures
A/@ [PCPS 76,161-172]

1380. Uniform asymptotic expansions(integrals with n saddle-pts)
A/@ [PCPS 76,211-231 (?URSELL)]

1381. Approximate differentiability
δ/A/@ [PCPS 76,33-43]

1382. Extensions of Holder's inequality
A/@ [PLMS 11,311-326]

1383. Change of order of fractional integration
δ   [PLMS 11,213-238]

1384. Generalizations of Chebychev's inequality
A/@ [PLMS XIII #51,385-412]

1385. Topologies on finite sets (combinatorics)
δ/A/@ [HOFFMANN/McCARTHY,eds.,Proc.10th SE Conf.Combin.,1979]

1386. Graph-forms and energy (Hydrocarbons)
δ/A [MR 81b #05079]

1387. The number of 'numerical evaluations' for iterated powers
δ   [BAMS 78,1092-1103*]

1388. Stable topologies
δ   [BAMS 75,493-498*]

1389. A Mathematical Theory of Systems Engineering
δ/A [WYEMORE,A.W.(Wiley,1967;pp353)]

1390. Invertible topological spaces
δ/A [BAMS:68,959-965;70,181-183;71,533-534;75,357-362,378-]

1391. Coincidence of f,g(TS-->HS)equal on a dense subset ...
$\delta/@$ [BAMS 75,390-391 (HS:=Hausdorff S)]

1392. Decomposability of positive functions on R^n
$\delta/@$ [BAMS 75,350-357]

1393. Total curvature and critical path theory (?OR!)
δ/A [BAMS 77,475-485]

1394. Variants of Radon inversion
$\delta/@$ [AMM 89,371-384*]

1395. Purely-algebraic formulation of 'calculus'
δ/A [AMM,89,362-370* ;KOCK,A.,LMS Lec.N.(CUP)*]

1396. Domains of NONdifferentiability of continuous functions
δ [BAMS 73,57-61]

1397. Use of Category Theory for finite-state machines
δ/A [LN Econ./Math.Syst.#115]

1398. Solutions of $(?)(d_x^n y)(d_y^n x) = 1$ (!!)
δ

 [BAMS 74,578-580]

1399. Inversion of the van der Monde matrix
δ

[BAMS 74,571-574]

1400. Periodic sequences of derived sets of M in S
δ   [BAMS 74,555-556]

1401. Topological props.of point-sets def.by finite automata
δ/A   [BAMS 74,539-542]

1402. { Counter-examples in TVSs }
δ/@   [LNM #936 (KHALEELULLA,1982)]

1403. Generalizations of 'uniform convergence'
A/@   [Colloq.J.BOLYAI,#8 (Csaszar,ed.),327-340*]

1404. Directional structures
δ/A   [Colloq.J.BOLYAI #8,187-211*]

1405. Variational techniques for planar elasticity
δ/A   [J.IMA 1,76-100*]

1406. Estimation of various types of probabilities
δ/A   [J.IMA 2,364-383*]

1407. Function-space forms of the Liapounov method
δ/A/@   [J.IMA 4,78-93]

1408. A general fp-theorem VIA Eilenberg/Steenrod homol.axioms
δ/@   [AMM 75,523-525]

1409. Conditions for a TS to be statistically metrizable
A/@ [Genl.Topol.Appl.9,233-237*]

1410. Maximal measure of POSETS with bounded chain measures
δ/A [Studies Appl.M 66,91-93*]

1411. Space-filling curves and stochastic independence
δ/A/A [AMM 88,426-432*]

1412. Convergence for iterates of self-maps of discs
A/@ [AMM 88,396-407*]

1413. Commuting,open cts.self-maps with common fp(s)
δ/A [AMM 71(6) (COHEN,H.,et al.)*;AMM 73,735-738*]

1414. Radii of univalency ...
δ [AMM 78,1031- *]

1415. Complex-integral proof of the Cayley-Hamilton theorem
δ/A [AMM 78,1003-1004*]

1416. Convergence of SUM({u(n)}),with u(n+1)=f(u(n))
δ/A [AMM 71,994-998*]

1417. Polygonal dissections ...
δ [AMM 71,1077-1095*]

1418. Uniform ctty. of products of classes of u.-cts functions
δ [AMM 72,20-28*]

1419. Elementary proof of 'No Retraction Theorem' in R^n
   δ [AMM 88,264-288*]

1420. ODEs with simply-constructible solutions
   δ [AMM 89,198-208* (?SAM)]

1421. Expository proof of the (CV) Uniformization Theorem
   δ [AMM 88,574-592*]

1422. A MS-model for (non-)euclidean geometries
   δ/A [AMM 74,673-677*]

1423. Unified treatment of Hardy-Littlewood Maximal Theorems
   δ/A [AMM 74,648-660*]

1424. Flow-invariance for sets(relative to ODEs)
   $\delta/@$ [AMM 79,740-747*]

1425. 'Horizontal chord theorems'
   δ [AMM 79,468-475*]

1426. Geom. constructions for cts.-nowhere differentiable fns
   δ/A [AMM 73,515-519*;AMM 68,653-655*]

1427. Convergence of SUM{ $f^{(n)}(x)$ } (n-th iterates)
   δ [AMM 75,964-968*]

1428. Sin(piA)/(piA) >= (1-AA)/(1+AA) ,A real
   A

[AMM 76,1153-1154*]

1429. Mathematical cartography
δ/A [AMM 76,1101-1112*]

1430. Characterization of SEMIMETRIZABILITY
δ/A/@ [AMM 73,361-365*]

1431. TS-definitions of 'simple connectedness'
δ [AMM 74,1117-1120*]

1432. Radon-Nikodym theorem as a result in probability theory
δ/A [AMM 85,155-165*]

1433. A 'triangle inequality' for (x(1),...,x(n),z)
A/@ [AMM 85,105-106*]

1434. Frechet differentials in { f:R-->R }
δ/A/@ [AMM 81,1006-1008*]

1435. 'Instability' of the intermediate-value prop. under +
δ/A/A [AMM 81,995-997*]

1436. Generalizations of Inv.-Fn.Thm. (strong differentiability)
δ/A/@ [AMM 81,969-980*]

1437. Multiple-integral reprns. of initial-value problems
δ/@/A [AMM 84,201-204*;1-D stat.mech.(Baxter)*]

1438. Nonextendability of 'order of differnetiability'
  δ   [AMM 84,162-167*]

1439. A general 'product integral inequality'
  A/@ [AMM 83,723-725*]

1440. Succint construction of the completion of a MS
  δ   [AMM 75,62-63*]

1441. Explicit representation of M^n for arbitrary matrices M of order 2
  δ   [AMM 75,57-58* (?SAM)]

1442. { Comprehensive(classical)'mathematical dictionary'
  δ   [NAAS/SCHMID (Teubner/Pergamon,1961/1962);Inst.]

1443. { Encyclopedic Dictionary of Applied Mathematics (Engrng.)
  δ   { [SNEDDON,I.N.,ed. (Pergamon P,c1980;pp808]

1444. Circulants over abelian groups
  δ/@ [AMM 73,207-208*]

1445. A cts. nowhere-diffbl.fn.vanishing on a set of measure >0
  δ   [AMM 73,166-168*]

1446. Limits of functions involving binomial coefficients
  δ/A [AMM 73,162-165*]

1447. The number of topologies on a finite set
S/A   [AMM:73,154-157*;75,739-741*]

1448. Matrix-theoretic proof of the AM/GM inequality
A/@   [AMM 74,305-306*]

1449. Indexed systems of neighbourhoods in general TSs
S/A   [AMM:68,886-893* ;74,283-284*]

1450. A generalization of the Perron-Frobenius theorem
A/@   [AMM 74,274-276*]

1451. Thermodynamic laws proven VIA use of generalized means
S/A/A   [AMM 74,271-274*]

1452. Volume of an n-simplex in R^n
S   [AMM 73,299-301*]

1453. 'Stability' for 'perturbed arithmetic functions'
S/A/A   [AMM 73,265-268*]

1454. A 'CV Rolle's Theorem'
S/@   [AMM 74,452-453*]

1455. Derivation of matrix reprns. from finite group tables
S   [AMM 74,430-432*]

1456. Near-homogeneity in 2-manifolds
S/A/@   [AMM 74,423-426*]

1457. Curvature-characterization of ellipsoids
  δ   [AMM 74,416-418*]

1458. Characterizations of the exponential distribution(stats.)
  δ   [AMM 74,414-416*; Bk*(?Wiley);Et al./LINNIK (Inst.)]

1459. Countability of {extreme points} in T_2 - spaces
  δ   [AMM 74,405-406*]

1460. Almost-measures of symmetry
  A/@   [AMM 74,820-823*]

1461. Conditions for permutation-invariance of series summation
  δ/A   [AMM: 80,317-318*;73,822-828*;PAMS 6(1955)563-564]

1462. Reachability--for vector addition systems
  δ/A   [AMM 80,292-295*;Int J Syst.Sci.8,321-338(dynam syst)]

1463. Algebras derived from graphs
  δ/A   [AMM 80,288-289*]

1464. Propagation of inequalities beween analytic functions
  δ/A   [AMM 73,555-556* ;AMM 84,1-12*]

1465. Extensions of the 'Integral Test' for series convergence
  δ/A   [AMM: 73,521-525* ;71,294-295*]

1466. Degree of approximation of f(x,y) by g(x) + h(y)
  A   [AMM 72,1101-1103*]

1467. Strong connectedness in TSs
  S   [AMM 72,1098-1101*]

1468. FOLDING of one subset along another subset in a MS
  S/@ [AMM 72,1094-1096*]

1469. Transform operators for difference-equations
  S   [AMM 72,747-750*]

1470. A 'vector mean-value theorem'
  S/A/@ [AMM 79,381-383*]

1471. Bibliography of schlicht functions
  S   [BERNADI,S.D. (NYU,1966;pp157,1694 titles)]

1472. Liouville- numbers and thick/thin sets
  S/A [AMM 83,648-650*]

1473. Sets of periods of real-valued periodic functions
  S   [AMM 73,761-762*]

1474. Uniform lim. of multiple balayage of integral wrt measures
  S/A [AMM 73,733-735*]

1475. Symbolic algorithms for higher curvatures in R(n)
  S/@ [AMM 73,699-704*]

1476. Countable set with limit-pt to wh. no subseq. converges
      δ   [AMM 75,1098-1099*]

1477. 'Converse' of the Divergence Theorem
      δ   [AMM 71,442-443*]

1478. Inverses of van der Monde matrices
      δ   [AMM 71,410-411*]

1479. BV and uniform convergence
      δ/A [AMM 71,537-539*]

1480. Metrizability and invertible TSs
      A/@ [AMM 71,533-534*]

1481. The number of partitions of a set
      δ   [AMM 71,498-503*]

1482. 'Differentiation procedures' for proofs of inequalities
      δ/A [AMM 76,543-546*]

1483. General formulation of 'change of vars.' for 1-D integrals
      δ   [AMM 76,514-519*]

1484. Measures in denumerable spaces
      δ/A/@ [AMM 76,494-502*]

1485. Quasi-continuous functions on ordered sets
δ/A/@ [AMM 76,489-494*]

1486. Functional-anal. proofs of basic function- theory results
δ/A [AMM 76,483-489*]

1487. 'Functional similarity' in R(n) (composition)
δ [AMM 76,627-632*]

1488. TSs with minimum neighbourhoods
A/@ [AMM 76,616-627*]

1489. Survey of 'geometric conditions' for the fp-property
δ/A [AMM 76,119-132*]

1490. 'Geometric characterizations' of 'analytic functions'
δ/A [AMM 71:257-264*;265-277*]

1491. Characterizations of sets of measurable functions
δ/A [AMM 84,455-458*]

1492. Consistent definitions of iterated powers
δ [AMM 81,643-647*]

1493. A converse of the Jordan Curve Theorem
δ [AMM 81,636-639*]

1494. Variants of Kakeya's Problem
δ [AMM 81,582-592*]

1495. BOUNDS on functions of matrices
δ/A [AMM 74,920-926*]

1496. Various definitions of 'functional independence'
δ [AMM 74,911-920*]

1497. Simple 'enumeration functions ' for the Rationals
δ [AMM 72,1013-1014*]

1498. Extensions of the B-W Theorem
δ/@ [AMM 72,1007-1012*]

1499. 'Normal metrics'(exhibit disjointness ...)
δ/A [AMM 72,998-1001*]

1500. Measures of non-normality for matrices
δ/A [AMM 72,995-996*]

1501. Rigorous treatment of Dimensional Analysis
δ [AMM 72,965-969*]

1502. Proofs/extensions of Young's Inequality
δ/A/@ [AMM:78,781-783;77,?603-609;81,760-761]*

1503. Functional-analytic proof of Rouche's theorem
δ/A [AMM 78,770-771*]

1504. Variants/extensions of the Banach fp-theorem
δ/A/@ [AMM 76,405-408*]

1505. 'Linearization' in rings/algebras
      δ/∂ [AMM 76,348-355*]

1506. Commuting mappings lacking commuting extension
      δ   [AMM 74,13-184*]

1507. Metrics on the power-set of a finite set
      δ/A [AMM 74,171-173*]

1508. Interesting 'offshoots' of the Fundamental Thm. of Algebra
      δ/A [AMM 71,180-185*;AMM 64,582-585*]

1509. Separation spaces as precursors of proximity spaces
      δ/A [AMM 71,158-161*]

1510. Strongly-continuous functions ...
      δ   [AMM 74,166-168*]

1511. Additivity,invertibility,and extendability in TSs
      δ/∂ [AMM 74,148-152*]

1512. Nonuniqueness of moment sequences (examples)
      δ   [AMM 72,302-303*]

1513. Generalized mean-value theorems
      δ/A/∂ [AMM:62,226-232*;72,300-301*]

1514. BOUNDS on zeros of polynomials VIA Gerschgorin's thm.
S/A [AMM 72,292-295*]

1515. Lambda-numbers of matrices(vibrations)
S [AMM 72,260-264*] / bk* (K. Lancaster)

1516. Characterizations of the 'geometric distribution'
S [AMM 72,256-260*]

1517. 'Order' in 'algebra',and in 'logic'
S/A [AMM 72.386-390*]

1518. Characterization of extreme doubly-stochastic measures
S/A/@ [AMM 72,379-382*]

1519. Optimal inequalities for probabilities of 'fns.of events'
A/@ [AMM 72,343-359*]

1520. An 'integral test' for convergence of recursive series
S/A [AMM 82,827-829*]

1521. Algebraic characterizations of 'limit operations'
S [AMM 82,825-827*]

1522. Characterizations of 'normal families of functions'
S/@ [AMM 82,813-817*]

1523. Limits of composite functions ... on TSs
S/A/@ [Amm 84,49-52*]

1524. Continued fractions over algebraic-number fields
S/@ [AMM 84,37-39*]

1525. A P-S that is uniformly, but not absolutely, convergent
S/A [AMM 82,507-510*]

1526. A converse of Banach's fp-theorem
S/A [AMM:75,775-776*;74,436-437*;PAMS 18(1967)287-289]

1527. Approximate bases in HSs
S/A/@ [AMM 75,750- *]

1528. Rearrangement-invariant series over Hausdorff-topol. grps
S/@ [AMM 75,729-731*]

1529. Algebraic analogues of CONVEXITY
S/A [AMM 75,879-880*]

1530. Dense topologies on various classes of spaces
S/A/@ [AMM 75,847-852*]

1531. Basic questions involving 'separation-of-variables'
S   [AMM 75,844-847*]

1532. Lower BOUNDS for mod(SUM { z : z in B })
A   [AMM 81,787-788*]

1533. Lebesgue-decompositions of probability distributions (EXLs)
S   [AMM 76,297-299*]

1534. A cts,strictly-incr.fn.whose derivative vanishes pp
   δ   [AMM 76,295-297*]

1535. Explicit representations of exp{tA},for matrices A
   δ   [AMM:76,289-292*;74,1200-1204*;GOLUB/van LOAN (bk)]

1536. The expected volume of a random simplex
   δ/A [AMM 76,286-288*]

1537. A 'Krein-Milman Theorem' for POSETS
   δ/@ [AMM:76(3)*;61,223-233;PAMS 51(1942)569-582]

1538. Means of one function relative to another function
   δ   [AMM 76,252-261*]

1539. Equivalence of 'compactness',and,'finiteness',in topology
   δ   [AMM 75,178-180*]

1540. Is the fixed-point property preserved by U (?union oprn)?
   δ   [AMM 75,152-156*]

1541. The 'mathematics of physical quantities'(models,...)
   δ   [AMM 75,115-138* (Pt I);WHITNEY,H.;Pt II* :=#1554]

1542. A generalization of Cauchy's inequality
   A/@ [AMM 76,815-816*]

1543. Characterizations of 'convergence in probability'
   δ/A/@ [AMM 76,813-814*]

1544. Uses of HOMOTOPY in function theory
∫   [AMM 76,778-787*]

1545. 'Intersection' and 'covering' properties for convex sets
∫/A   [AMM 76,753-766*]

1546. General formulae for the n-th derivative of {1/f(x)}
∫   [AMM 74,1239-1240*]

1547. Characterizations of 'almost-uniform convergence'
∫/A/@ [AMM 74,1230-1231*]

1548. The 'K-product' for arithmetic functions
∫   [AMM 74,1216-1217*;Canad.J M 17(1965)970-976]

1549. Periods of meas.fns/constrn.of Stone-Cech compactification
∫/A/@ [AMM 71,891-893*]

1550. Momotone functions generated by ergodic sequences
∫/A/A [AMM 75,594-601*]

1551. Minkowski's 'geometric inequality' for 'mixed areas'
A/@   [AMM 75,581-593*]

1552. Characterizations of 'well-chained MSs'
∫   [AMM 75,273-275*]

1553. Periodic maps and commuting maps
δ   [AMM 75,265-268*]

1554. The 'mathematics of physical quantities' (continuation)
δ   [AMM 75 227-256 (Pt II);WHITNEY,H. ;Pt I := #1541]

1555. Fixed-point theorems for operators on subsets
A/@   [AMM 73,1022- *]

1556. Schnirelmann densities for n-dimensional sets
A/@   [AMM 73,976-979*]

1557. INTRINSIC METRICS:construction procedures/background
δ/A/@   [AMM 73,937-950*]

1558. Algorithmic solution of systems of Boolean equations
δ   [AMM 73,29-35*]

1559. Theory/APPLICATIONS of abstract integration
δ/@   [SEGAL,I./KUNZE,Integrals and Oprs*(McG-H,1968,pp308)]

1560. Numerical studies of quadratic,area-preserving maps
δ/A   [QAM 27,291-312*]

1561. Distance-matrices of graphs (realizability ...)
δ   [QAM 22,305-317*]

1562, 'Normal modes' in NONLINEAR systems
δ/@   [QAM:22,217-234;24,177-193 (?ROSENBERG,R.M.)]

1563. Use of topological spaces in INTERVAL ANALYSIS
δ/A/A   [RAAG Memoirs;bks.by MOORE,R.*-]

1564. Solution techniques for WORD PROBLEMS in groups
δ   [MAGNUS,W.,et al.,Combinatorial Grp Th (bk)*]

1565. Fourier-series proof of the Euler-Lagrange equations
δ   [AMM 74,587-588* (Calculus of variations)]

1566. General discussion of algebraic structures
δ   [ROSENFELD,A.(Holden-Day,1968,pp285 (uncommon topics))]

1567. Converse of Rouche's theorem
δ   [AMM 89,302-305*]

1568. De-coupling of pairs of PDEs
δ   [QAM 27,87-104 (?SAM)]

1569. Foundations of pattern analysis
δ/A   [QAM 27,1-55*;bks,Vols I-III (Springer)*- all,GRENANDER]

1570. Iterative solution of Wiener-Hopf integral equations
δ/A   [QAM 20,341-352* (?SAM)]

1571. Theory of NONLINEAR NETWORKS
δ/A/A   [QAM:22,1-33;22,51-104 (Pts I,II:?MOSER,J.)]

1572. Approximation techniques in SHELL-THEORY
A/A   [GOL'DENVEIZER (Pergamon P,1961;pp658) ;?Inst.]

1573. PRESENTATIONS of group extensions
    S   [LMS Lec.N #42 (?JOHNSON)*-]

1574. ESTIMATES of orders of finite groups
    S/A [GORENSTEIN,D.,bk(Harper & Row,1968)*,Secs.9.1,9.4]

1575. Vector-valued Nevanlinna theory
    S/@ [ZIEGLER,H.J.W.(Pitman Res.N #73]

1576. Symbolic calculus for ASYMPTOTICS
    A/S [GUILLEMIN/STERNBERG, bk(AMS,1977);cf.,MASLOV,bk*]

1577. Decomposition of MUTIVARIATE PROBABILITIES
    S   [CUPPENS,R.(AP,1975,pp244)]

1578. Approximation of CONVOLUTIONS by Gaussians in R(k)
    A/@ [Th.Prob.Appl.(USSR)22(1977):two papers*]

1579. Stochastic calculus for PLANAR MARTINGALES
    S/@ [Ann.Prob.4(1976)570-581?* (and refs.)]

1580. BOOLEAN metric spaces
    S/A [MR 23A #A179 (p31)*]

1581. Differential calculus in LCTVSs
    S/@ [LNM #417(KELLER)*]

1582. BVs of analytic functions/generalized functions
δ/@ [BELTRAMI/WOHLERS (AP,1966;pp116)   Inst.   ]

1583. NEURONS as Lie-group germs/products
δ/@/A [QAM 25,423-440*]

1584. POSITIVE OPERATORS in BSs
δ/A/@ [J.Math.Mech.8(1959)907-937 (KARLIN,S.)?*   ]

1585. COMPUTABLE algebra
δ/A [TAMS 95(1960)341-360?*;MR 22A #4639*]

1586. SPATIAL matrices (all elements have p subscripts ...)
δ/@ [SOKOLOV,N.P.,bk (Moscow,1960)]

1587. A 'Stone -Weierstrass theorem' for PROXIMITY SPACES
δ/A/A [Fund.Mat.47(1959)205-213;CECH:TSs(Wiley,1966)*]

1588. N-dimensional analogues of Cauchy's integral theorem
δ/A [J.Austral.M S 1(1959/61)171-202]

1589. The 'Riemann Hypothesis'(RH) for function fields
δ/@ [HASSE,H.,U.Madrid Publ.Soc.Mat.Fac.(1957)]

1590. A generalization of Laplace transforms
δ/@ [Rend.Circ.Mat.Polermo(2)6(1957)325-333?*]

1591. CONVEXITY in TSs
δ/@ [Colloq.Mat.6(1958)283-286;Canad.J M 9,511-514;Indag.Mat.
       21(1959)36-38; MR 21 ##3828,3829]

1592. Convexity results derived VIA integration over Unit Sphere
δ   [AMM 90,690-693*]

1593. Functions which 'parametrize means'
δ/A [AMM 90,677-683*]

1594. Function-space ANALOGUES of theorems for sequence-spaces
δ/A [QJM 11(1960)310-320?*; MR 22A #12371]

1595. 'Representation by functions of fewer variables'
δ   [Mat.Sbornik 48(90) (1959) 3-74;MR 22A ##12191,12192]

1596. Approximation of FINITE systems by INFINITE systems
δ/A [GHOSAL,A.bk(G&B,1978)*,p96]

1597. Fourier transforms of measures supported by CANTOR SETS
δ   [SALEM,R.,Alg.Nos.&Fourier Anal.(Heath,1963)*,Ch.4]

1598. Information-theoretic proof of the Central Limit Theorem
δ/A/A [Th.Prob.Appl.4,288-299*]

1599. REFORMULATION of BV- problems as IV-problems
δ   [JACM 18(1971)594-602*]

1600. GENERATION of differentiable surfaces by INTERPOLATION
δ/A [JACM 18,576-585*]

1601. Equilibration/equal column-norms for matrices
  δ   [JACM 18,566-572*]

1602. Eigenvalue problems and 'recognition of 0'
  δ   [JACM 18,559-565*]

1603. Algebraic simplification VIA HASHING
  δ   [JACM 18,549-558*]

1604. Analysis in 'computable number fields'
  δ/A [JACM 15,275-299*;Bk :ABERTH,O.]

1605. Parallel matrix formulation of the FFT
  δ/@ [JACM 15,252-264*]

1606. Formal analysis of functional contour maps
  δ   [JACM 15,205-220]

1607. Perspective representations for fns. of two variables
  δ   [JACM 15,193-204*]

1608. Parallel reformulation of DYNAMIC PROGRAMMING
  δ/@ [JACM 15,176-192*]

1609. Resultant procedures /Graeffe process(roots of alg.eqs.)
  δ   [JACM 7,346-386* (?STEIFEL,E.)]

1610. Minimization of Boolean functions
  δ/A [JACM: 7,299-310*;11,283-295*;IRE Trans.(9/56),126-132]

1611. Definitions of EFFICIENCY for algorithms
δ   [JACM 17,708-714*]

1612. ACCELERATION for general algorithms:LIMITATIONS
δ/A [JACM 18,290-305*]

1613. Characterization of SWITCHING FUNCTIONS(Chow parameters)
δ/@ [JACM 18,265-289*]

1614. BOUNDS on matrix condition-numbers
δ/A [JACM 21,514-524*]

1615. 'Minimal programs' of prescribed specification
δ/A [JACM 21,436-445*]

1616. Information -theoretic criteria for thm.-proof complexity
δ/A/@ [JACM 21,403-424*]

1617. Group-theoretic PARTITIONING ALGORITHMS (matrix inversion)
δ   [JACM 16,302-314*]

1618. Invariant aspects of POLYADIC AUTOMATA
δ/A [JACM 16,302-314*]

1619. Topological aspects of FINITE-DIFFERENCE techniques
δ/A [JACM 18,63-74*]

1620. Experiments in interactive THEOREM-PROVING(alg./anal.)
δ   [JACM 16,49-62*]

1621. Statistical models for the solution of Fredholm int.eqs.
δ/A   [JACM JACM 21,1-5*]

1622. Abstract definitions of COMPUTER
δ/@   [JACM 13,226-235*]

1623. 'Diffusion approximations' in QUEUEING THEORY
δ/A/A [JACM:21,316-328*;21,459-469*]

1624. 'Polarized-distance techniques' in INFORMATION RETRIEVAL
A/A   [JACM 21,233-245*]

1625. Topological properties of quantized spaces(CS)
δ/A   [JACM 20,439-455*]

1626. Topological properties related to SOLVABILITY(numer.anal.)
δ/@/A [JACM 20,399-408*]

1627. Existence of arbitrarily large COMPLEXITY GAPS
δ/A/@ [JACM 19,158-174*]

1628. Spectral norms of iterative processes
δ/A   [JACM 6,494-505*]

1629. Canonical forms for Boolean functions
δ   [JACM 6,245-258*]

1630. AUTOMATH (Survey)
δ   [LNM #125*- ;ALSO:many Eindhoven Reps.;bk*(REZUS)]

1631. MODELS -- in terms of structure-preserving morphisms
δ   [JACM 19,742-764*]

1632. Generation of minimal trees with the STEINER TOPOLOGY
δ/A/@ [JACM 19,699-711*]

1633. Computational representations of LIE ALGEBRAS
δ   [JACM 19,577-589*]

1634. ⎰Overview of COMPUTATIONAL COMPLEXITY
δ/A   ⎱[JACM 18,444-475;GAREY/JOHNSON;PAPDEMITRIOU/STEIGLITZ,bks]

1635. Error analysis of direct matrix inversion
δ/A   [JACM 8,281-331(?WILKINSON;&bk*)]

1636. Automata minimizing the effects of INPUT ERROR
δ/A   [JACM 11,338-351*]

1637. Factorization of COVARIANCE functions
δ/@   [JACM 23,310-316*]

1638. BOUNDS for parallel evaluation(ratl.expr./recurrences0
δ/A   [JACM 23,252-261*]

1639. BOUNDARY-CONTRACTION methods for the Laplace equation
δ/A   [JACM JACM 6,227-235*]

1640. Parallel scheme for matrix inversion
δ/@ [JACM 14,757-764*]

1641. Applications of quasi-euclidean distance functions
A/@ [JACM 15,600-624*]

1642. Statistical criteria for 'theorem status'
δ/@ [JACM 19,347-365*]

1643. Comparisons of POST and BOOLEAN algebras
δ [JACM 21,680-696*]

1644. Use of the CHURCH-ROSSER property in program optimization
δ/A/@ [JACM 21,671-679*]

1645. UNDECIDABILITY results for classes of singular integrals
δ/@/A [JACM 21,586-589*]

1646. DECOMPOSITION of arbitrary finite automata
δ/@/A [JACM 15,135-158*]

1647. Some ASYMPTOTIC ESTIMATES in automata theory
A/A [JACM 13,151-157*]

1648. Direct products of automata ...
δ/@ [JACM 12,187-195*]

1649. INTERRELATIONS: Turing machines/finite automata/neural nets
δ/@/A [JACM 8,467-475*]

1650. PERCEPTRONS (pattern recognition ...)
δ/A   [JACM 8,1-20*]

1651. Automatic consistency/reduction proc. for 'alg.PDEs'
δ/A/@ [JACM 14,45-62*]

1652. ANALOGUES of connectedness/n-connectedness(digital plots)
δ/A   [JACM 17,146-160*]

1653. Efficient generation of CONVEX POLYTOPES in R(n)
δ   [JACM 17,78-86*]

1654. Set-formation relative to partition(s) (Complexity)
δ   [JACM 22,215-225*]

1655. Continuity of ROUND-OFF error in DATA ...
δ/A   [JACM 22,512-521*]

1656. BOUNDS for parallel evaluation of classes of expressions
A/@   [JACM 22,477-492*]

1657. Canonical forms and simplification
δ/@   [JACM 17,385-396*]

1658. 'Limit operations' on digital plots ...
δ/A/@ [JACM 17,348-360*]

1659. Local/global complexity of planar geometrical properties
δ/A

[JACM 17,339-347;Proc.Sympos.Appl.Math 19(AMS),176-218]

1660. Matrices formed as KRONECKER PRODUCTS
δ   [JACM 17,260-268*]

1661. Automata and computable probability measures
δ/A [JACM 17,241-259*]

1662. Connectivity and reversibility in automata
δ/A [JACM 17,231-240*]

1663. Efficient solution of RECURRENCE relations
δ   [JACM 14,563-590*]

1664. Approximation procedures for FREE-BOUNDARY problems
δ/A [JACM 17,397-411*]

1665. Information-theoretic definitions of PROGRAM SIZE
δ/@ [JACM 22,329-340*]

1666. Electrical ANALOGUES of stochastic models
δ/@/A [MPCPS 80,145-151*]

1667. Pre-ordered SYNTOPOGENOUS spaces
δ/A/@ [MPCPS 80,71-79*]

1668. RIEMANN-summability
δ/A [MPCPS 75,83-94*]

1669. Probability-density forms of the Central Limit theorem
S/@ [MPCPS 75,365-381*]

1670. Metrics/tauberian conditions for sums of RVs
S/@/A [MPCPS 75,361-364*]

1671. Steepest descents:n parameters,coincident saddle-points
S/A/@ [MPCPS 72,49-65*]

1672. Optimal recursion VIA fixed points ...
S/A/@ [CACM 20,824-831*]

1673. Symbolic computation for GROUPS (Survey)
S/@ [CACM 12,3-12*]

1674. Time-estimates/means for sequences of matrix products
S/A [CACM 16,22-26*]

1675. Generation of 'Ising configurations' (stat.mech.)
S/A [CACM 12,562- *]

1676. STABILITY/ACCURACY analyses for stiff ODEs
S/A [CACM 18,53-56*]

1677. Algebraic formulations of STRUCTURED PROGRAMMING
S/@ [CACM 18,43-48*]

1678. Conjugate gradients and PSUEDO-INVERSES (of operators)
S/A/A [CACM 18,40-42*]

1679. Elementary divisors of Kronecker products of matrices
    δ   [CACM 18,36-39*]

1680. Eigen-value PERTURBATIONS for non-normal matrices
    δ/A [CACM 18,30-36*]

1681. 'Hadamard condition- numbers' for matrices
    δ/A [CACM 18,25-29*]

1682. Stability of Gauss/Jordan elimination with pivoting
    δ/A [CACM 18,20-24*]

1683. Computation of 'leminscate constants'(elliptic integrals)
    δ   [CACM 18,14-19*]

1684. Norms and POSITIVITY for matrices
    δ/A [CACM 18,9-13*]

1685. 'Average behaviour' of selection algorithms
    δ/A/@ [CACM 18,165-172*]

1686. Computer reliability/markov processes (SAM)
    δ/A [CACM 21,586-591* ; bk*]

1687. Partial ordering of events in DISTRIBUTED computer syst.
    δ/A/@ [CACM 21,558-565*]

1688. 'Effective product-order'for multiple matrix products
    δ/A [CACM 21,544-549*]

1689. BOUNDS for complexity of measure{U[b(j),c(j)]}
δ/A [CACM 21,540-544*]

1690. Generation of 'arbitrary 3-D surfaces'
$\delta/@$ [CACM:20,703-712;19,454-460;19,555-563]*

1691. Surface-approximation VIA triangulation
δ/A [CACM 20,693-702*]

1692. VERIFICATION of structured programs
δ [CACM 20,271-283*]

1693. Recursive techniques of program verification
δ [CACM 20,209-222*]

1694. Logic and programming languages
δ [CACM 20,634-641* (SCOTT,D.,Turing Lec.)]

1695. Complexity of computation
δ/A [CACM:20,624-633;21,231- (Correc.)(RABIN,M.,Turing Lec)]*

1696. Integral-equation representations(parabolic Cauchy probl.)
$\delta/@$ [CACM 15,1050-1052*]

1697. Complete calculi for matrices
$\delta/@$ [CACM:13,223-237;15,1033-1039]*

1698. Flexible data structures (SETL)
   δ   [CACM 18,722-728*]

1699. REDUCTION:for proofs of properties of parallel programs
   δ/@ [CACM 18,717-721*]

1700. Automatic PROGRAM-SYSTHESIS
   δ   [CACM 14,151-165*]

1701. Formal verification for parallel programs
   δ   [CACM:19,279-284;19,371-384]*

1702. Translation:LANGUAGES --> FLOW-CHARTS (for evaluation)
   δ/A [CACM,15,967-973*]

1703. Recursive data-structures in APL
   δ   [CACM 22,79-96*]

1704. Visual- display facilities for n-D objects
   δ/@ [CACM 10,469-473*]

1705. Error analysis for rational computations
   δ/A [CACM 15,813-817*]

1706. Formal(anl./geom.)approaches to INFORMATION RETRIEVAL
   δ/A [CACM:15,802-808;18,464-470;18,613-620]*

1707. Algorithms for 'integral properties of solids'
   δ/A [CACM:25,635-641;25,642-650]*

1708. (i) Computer generation of 'stochastic objects'
δ/A/A [CACM 25,371-384*]
       (ii) Various PAPERS ON SAM
       [Proc.2nd SIGSAM Sympos.:JACM 14(8)*]

1709. FUNCTIONAL PROGRAMMING
  δ   [CACM 21,613-641 (BACKUS,?J.;Turing Lec.)]

1710. Hidden-part elimination in IMAGE RECONSTRUCTION
δ/A/@ [CACM 12,206-211*]

1711. NORMS for numerical integration of analytic functions
 δ/A   [CACM 12,268-270*]

1712. Tutorial on DENOTATIONAL SEMANTICS
 δ/@   [CACM 19,437-453*;bk(STOY,J.)*]

1713. Languages for statements about ARRAYS
  δ   [CACM 22,290-298*]

1714. Formal verification VS proof
 δ/@   [CACM 22,271-280*]

1715. Algorithms for the GCD of n integers
  δ   [CACM:13,433-436;13,447-448]*

1716. Solution of the equation w exp{w} = x
 δ/A   [CACM 16,123-124*]

1717. Error-BOUNDS for the simple zeros of analytic functions
S/A   [CACM 16,101-104*]

1718. Languages for the COMMUNICATION of mathematical informn.
S/@/A [CACM:15,71-75;16,275-281]*

1719. INTERFERENCE between parallel processes
S/A   [CACM 15,427-437*]

1720. ALGORITHMS for the solution of sets of n nonlin. eqs
S/A/@ [CACM 10,726-729;SIAM J Num.Anal.3,650658;ORTEGA/(bk)]*

1721. Interference checks for moving apparatus
S/A   [CACM 22,3-9*]

1722. Convolution computations VIA the FFT (numerical)
S   [CACM 12,179-184*]

1723. Formal assembly(construction) of solids
S/A   [CACM 18,209-216*]

1724. Best one-sided approximations
S/A   [CACM 14,598-600*]

1725. Information retrieval(various schemes)
S   [CACM:13,67-73;14,593-597]*

1726. Implementations of the REMEZ ALGORITHM(Chebychev approx.)
S/A   [CACM 14,737-746*;papers* by BRIDGES,D.S.(constructive)]

1727. Ordering proc.for {+/- f^(n)}(x) ,f positive,monotonic
δ/A [CACM 15,43-46*]

1728. CS as 'empirical enquiry':symbols and search proc.
δ/@/A [CACM 19,113-126* (NEWELL,A./SIMON,H.;Turing Lec.)]

1729. Motion of 'hidden lines' under rotations(image analysis)
δ/A [CACM 15,245-252*]

1730. Logical analysis of programs
δ/@ [CACM 19,188-206*]

1731. LG:a language for analytical geometry
δ/A [CACM 19,182-188*]

1732. Use of MULTI-SETS in proofs of program termination
δ/@ [CACM 22,465-475*]

1733. Exhaustive generation of floor-plans (designs)
δ/A [CACM 24,813-825*]

1734. Hierarchical representations of curves
δ [CACM 24,310-321*]

1735. Algorithms for the representation of SYMMETRIC SUMS
δ [CACM:10,428-429;10,450-]*

1736. Contour MAPS for functions of 3 independent variables
δ/A

[CACM 10,425-428*]

1737. Natural-SPLINE interpolation
δ/A   [CACM:16,763-768;17,463-467]*

1738. Generation of ROOT-LOCUS DIAGRAMS
δ   [CACM 10,186-188*]

1739. TENSORS in various co-ordinate representations
δ   [CACM:9,864- ;10,183-186]*

1740. Interval-arithmetic error estimation in linear algebra
δ/A   [CACM 12,89-93(and refs.)*]

1741. Roster of PROGRAMMING LANGUAGES (1974/75) (annotated)
δ   [CACM 19,655-669*]

1742. Languages for formal PROBLEM-SPECIFICATION
δ/A   [CACM 20,931-935*]

1743. Interrelations of CORRECTNESS and SPECIFICATION (programs)
δ/@   [CACM 21,159-172*]

1744. Reconstruction of images by RELAXATION procedures
δ/A/A   [CACM 21,152-158*]

1745. Preservation of AVERAGE PROXIMITY in arrays
δ/A/A   [CACM 21,228-231*]

1746. Automatic(dynamic) choices of DATA STRUCTURES (CS)
  δ   [CACM 21,376-384*]

1747. Correctness-proofs VIA test-data techniques
  δ   [CACM 21,368-375*]

1748. Conversion of RECURSION into ITERATION
  δ/A   [CACM 20,434-439*]

1749. Enumeration of FINITE TOPOLOGIES
  δ   [CACM 10,295-297*]

1750. POSE:a problem-posing language
  δ/A   [CACM 10,279-285*]

1751. Numerical integration of functions with POLES
  δ/A   [CACM 10,239-243*]

1752. Approximation of harmonic functions over STARLIKE domains
  δ/A   [CACM 17,471-475*]

1753. INTERPOLATION using C-infinity functions
  δ/A   [CACM 17,476-479*]

1754. DECISION-TABLE representations of programs
  δ   [CACM:17,532-537;16,532-539;13,571-572]*

1755. A CONTINUED-FRACTION algorithm for integration of ODEs
  δ/A/@ [CACM 17,505-508*]

1756. Inductive techniques for proofs of properties of programs
𝛿/A [CACM 16,491-502*]

1757. Bi-variate CUBATURE ...
𝛿/A/@ [CACM 16,567-570*]

1758. Skew representations of the SYMMETRIC GROUP
𝛿/@ [CACM 16,571-572*]

1759. Cubic-spline solutions to classes of FUNCTIONAL-DIFFL eqs
𝛿/A/@ [CACM 16,635-637*]

1760. Approximate solution of WIENER-HOPF integral equations
𝛿/A [CACM 16,708-710*]

1761. Conversion between two orthogonal expansions
𝛿   [CACM 16,705-707*]

1762. Numbering systems for sets of COMBINATIONS (Combinatorics)
𝛿   [CACM 17,45-46(and refs.)*]

1763. M-set PARTITIONS of a given set ...
𝛿   [CACM 17,224-225*]

1764. Binary patterns RECONSTRUCTIBLE from projections
𝛿   [CACM 14,21-25*]

1765. WIDTH and ANGLE as 'descriptions'
𝛿   [CACM 14,15-20*]

1766. TRANSLATORS of programming languages(?compilers)(Survey)
   δ   [CACM 11,77-113*]

1767. Early CS texts/curriculum outlines:Bibliography
   δ   [CACM 11(3)*]

1768. Estimate of various CONSTANTS in linear OD-operators
   δ/A   [CACM 11,255-256*]

1769. Calcln of plane-curve intersections/ singular pts of ODEs
   δ/A   [CACM 11,502-506*]

1770. Convergence-improvement for IMPROPER INTEGRALS
   A   [CACM 11,497-502*]

1771. Computations involving PERMANENTS
   δ   [CACM 12,634- ;bk (MINC,M.;CUP)]*

1772. Minimum-length polygonal approxns. to digital contours
   δ/A/@   [CACM 13,41-47*]

1773. Generation of NUMBER-PARTITIONS (?Number theory)
   δ   [8,493- ;13,120-]*

1774. Generation of COLLISION-FREE PATHS among polyhedra
   δ/A   [CACM 22,560-570*]

1775. Identification of ERGODIC SUBCHAINS/TRANSIENTS
    δ   [CACM 11 619-621*]

1776. The no. of AUTOMORPHISMS of a single-generator automaton
    δ/A [CACM 13,574-575*]

1777. Generation of possible(bearable!?)furniture plans
    δ   [CACM 18,286-297*]

1778. Minimum-area-encasing rectangle/arbitrary closed curves
    δ/A [CACM 18,409-412*]

1779. Conditions for SIMILARITY of matrices A,B(B,positive)
    δ   [CACM 11,559-560*]

1780. Algorithm for PETROV-CLASSIFICATION(gravitational fields)
    δ   [CACM 21,715-717*]

1781. Recursion-induction(for properties of recursive fns)
    δ/@ [CACM 14,351-354*]

1782. OPTIMAL DETECTION of curves in noisey graphical plots
    δ/A/A [CACM 14,335-345*]

1783. PERTURBATION OF CONSTRAINTS (non-smooth optimization) #46
    δ/A/@ [USC #7931* (AUBIN,J-P)]

1784. Asymptotic properties of LARGE GAMES(approximations) #171
    δ/A [Stony Brook ##215,216*]

1785. Approximate formulation of DEs for HOMOTOPY PATHS
δ/A/@ [U.Chicago #7931*]

1786. Stabilization of GLOBAL NEWTON METHOD (soln of equations) #6
δ/A/@ [U.Chicago #7937*]

1787. DIMENSION-REDUCTION(cf,ignorable co-ords;center manifolds) #7
δ/@ [U.Chicago #7914*]

1788. Differentiable/variational props.of'Polya distribn fns #65
δ/A [Kobe U.(Japan) #59(KAWADA)*]

1789. Topological methods for CONSTRAINED OPTIMA   #64
δ/A/@ [U.Kansas #103*]

1790. Number and types of EQUILIBRIA for general ECONOMIES
δ/A/@ [CEPREMAP #7905*]

1791. Continuity of FIXED-POINTS /deformations of cts mappings
δ/A/@ [Summa Brasil.Math.4(1960)183-191?*]

1792. Approximation of STOCHASTIC INTEGRALS
δ/A/@ [U.Wisconsin #594*]

1793. Condns for 2nd-order difference eqs to be variational eqs #104
δ/A [USC #7716*]

1794. Partial ordering of prob.distributions VIA UTILITY fns #27
A/A [N-W U. #153*]

1795. Cyclical motion in dynamical economies #85
S/A   [N-W U.#334*]

1796. Measures for LEVEL CURVES of smooth functions #176
S/A/@ [Berkeley #IP-273*]

1797. Integrability of CHOICE-SETS ... #21
S/A   [Berkeley IP-266*]

1798. VALUES of nondifferentiable GAMES #140
S/A   [CORE (Louvain) #7941*]

1799. Limit- theorem for the CORE (simple proof) #8
S/A/A [CORE(Louvain) #?7915*]

1800. LIMITS of finite games: 'partition values' #100
S/A/A [CORE(Louvain) #7913*]

1801. Existence of EQUILIBRIA in competitive economies #53
S/A   [Berkeley IP-269 (DEBREU)*]

1802. VALUE-CONVERGENCE theorems #18
S/A/A [CORE(Louvain) #7942*]

1803. Generic existence of eqilibrium under UNCERTAINTY #131
S/A/A [Berkeley IP-265*]

1804. EXPECTATION-FORMATION under uncertainty   #49
   S/A   [Cornell #209*]

1805. Prediction of RESPONSE to output change in duopoly   #166
   S/A   [U.Michigan #C-16*]

1806. HOPF-BIFURCATION anal./equilibria(multi-sec.economies) #32
   S/A/A   [USC #7919*]

1807. Lagrangian soln./constrained convex optmn probs./BS   #50
   S/A/@/A   [USC #7921]

1808. Optimal infinite-horizon programs(unlimited time)   #51
   S/A/A   [USC #7925*]

1809. Power(to change status quo)for n-person games   #142
   S/A   [N-W U. #366*]

1810. Formal characterizations of COMMON KNOWLEDGE   #82
   S/A   [N-W U. #393*]

1811. Strategy-proof allocation:ADJUSTABILITY   #106
   S/A   [N-W U. #395*]

1812. Equivalence of CORE and COMPETITIVE allocations   #117
   S/A   [USC #7801*]

1813. MUTUAL convergence of COMPETITIVE and VALUE allocations #96
   S/A/A   [USC #7805]

1814. EXISTENCE-thms for n-valued DEs with constraint/ordering #146
δ/∂/A   [USC #7707*]

1815. CORE of economy as a set of unimprovable states #172
δ/A/A   [Berkeley IP-260*]

1816. EQUIVALENCE of variational probs to Cauchy systems #147
δ/A/A   [USC #7710*]

1817. Parameter-dependence of eqilibria ... #120
δ/A   [USC #7721*]

1818. (Baysian) diffusion to nonco-operative equilibrium #111
δ/A/A   [NYU #79-01*]

1819. Test-function criterion for ERGODICITY of Markov chains #24
δ/∂/A   [CORE(Louvain) #7912*]

1820. Optimal growth-paths in presence of multiple steady states #148
δ/A/A   [USC #7817*]

1821. AUCTIONS as games with incomplete information #4
δ/A/A   [U.Minnesota #79-110*]

1822. Approximate equivalence of 'myopic'/global- max. expectn #45
A/A   [(USC #7929*)]

1823. Augmented DE systems for economic analysis #46
δ/A   [USC #7931*]

1824. Allocation/group-incentive equilibria/the CORE
δ/A [N-W U. #329*]

1825. Non-co-operative equilibria charac. by 'price-taking' #62
A/A [Princeton #Econ 228*]

1826. Adjustments to random disturbances #103
A/A [HILDEBRAND/RADNER (Paper D1)*]

1827. A definition of DIVERSIFICATION #212
δ/A [UCB(Business) Repr.#7*]

1828. Extension of VALUE to nonatomic econs.(equiv.to CORE) #182
δ/@/A [Stanford #87*]

1829. Extension of DECOMPOSITION techn.to NONLINEAR programs #139
δ/@/A [CORE(Louvain) #7940*]

1830. Economic equilibria under uncertainty/CORE
δ/A/A [Paper (DYNKIN,E.)*]

1831. Generalized ('fuzzy') games(equivalence to the CORE) #7
δ/@/A [USC #7914*]

1832. Adaptation to 'quantized changes' with protection #181
δ/@/A [USC #7831*]

1833. Partially controllable manipulation of economies #136
δ/A/A [U.Minnesota #79-116*]

1834. RECOVERABILITY of the cardinal-utility function #193
δ/A [CORE(Louvain) #7945*]

1835. Sets with the 'global Lipschitz normal' property #160
δ/A/@ [CORE(Louvain) #7926*]

1836. Non-isolability of Walras equil. from unsatisfactory equil. #194
A/@ [UCB IP-272*]

1837. Multivariate asymptotics for structural coefficients #155
δ/A/A [UCB IP-254*]

1838. Continuity of PRICE MECHANISM in CONSUMABLES #175
δ/A [CORE(Louvain) #7931*]

1839. Uniqueness of MEAN-MAXIMIZERS(Inf.-diml. 'Sard theorem') #174
δ/@/A [UCB IP-263*]

1840. Sufficient conds. for DIFFERENTIABILITY of mean demand #11
δ/A/A [UCB #257*]

1841. Time-asymptotic measures of SYSTEM PERFORMANCE #177/#66
δ/A/A [UCB:##ORC 79-6,79-7]*

1842. Existence of temporary equilibria in exchange economies #209
δ/A/A [USC #7824*]

1843. STABILITY of temporary equilibria in exchange economies #209
S/A/A [USC #7824*]

1844. Equivalence of several notions of EQUILIBRIA #130
S/@/A [UCB #IP-270*]

1845. Computation of the DIAMETERS of permutation-groups
S/@/A [Paper (CMU/CS (2/83);DRISCOLL,J.R.;pp16)*]

1846. Inversion theorems/defns.&intersections of Lorenz-curves #118
S/A [ISER-117*]

1847. Variation-norms/values of nonatomic games/internal spaces
A/@/A [Stanford Econ #407*]

1848. Separably infinite (linear) programs #119
S/A/A [CCS-34 (Austin,Texas)*]

1849. Positive-recurrence criteria for m-diml Markov chains #207
S [CORE(Louvain) #7903*]

1850. Existence of VALUE on space of all scalar integrable games #208
S/A [CORE(Louvain) #7902*]

1851. (Stationary) fps of dissipative n-valued mappings #17
S/A/@ [USC #7712*]

1852. SMOOTHING of noisey data with thin-plate SPLINES #186
A/A [U.Wisconsin #557*]

1853. TOPOLOGY of preference spaces/social choice #185/#116
S/A/A [Adv.Math.(1979+;CHICHILNISKY);Columbia U. Papers:
       ##4,32*,35*,16,19,25]

1854. LOCATION of centers on a TREE with 'discontinuities' #145
S/A   [N-W U. #397*]

1855. ALMOST-OPTIMALITY of 'partial communication' in econ.struc. #128
A/A   [Harvard #572*]

1856. Games for 'socially-produced altruism' #3
S/A   [Stanford #255*]

1857. Diffuse trader charac. for uncertain exchange economies #105
A/A   [U.Penn. #79-08*]

1858. Equilibria under uncertainty/price information #66
A/A   [U.Penn. #79-07*]

1859. Strategy-proof allocation mechanisms #106/#145
S   [N-W U. ##395*,396*]

1860. Geometric EQUIVALENCE of two fp-representations #68
S/A/@ [N-W U. #365*]

1861. Equilibrium theory/capital theory (turnpike theorems) #15
A/A/@ [N-W U. #405*]

1862. Cyclical motion in stochastic/deterministic economies #85
S/A   [N-W U. #334*]

1863. RECOVERABILITY of the cardinal utility function #87/#135
S/A/A [Columbia U. #3*;CORE(Louvain) #7937*]

1864. Manipulation of game-values VIA preference MISreprtn. #136
S/A   [U.Minnesota #79-116*]

1865. Nash equilibrium and co-operation for 'fair games' #95
S/A/@ [Columbia U. #46*.(Games as fns between smooth manifolds)]

1866. Closed orbits in multi-sector econs.(Hopf bifurcations) #28
S/A/A [USC #7718*]

1867. Equivalence of variational problems to Cauchy systems #29
S/A/A [USC #7719*]

1868. Algorithms for Kth-shortest paths #204
S/A   [Kobe U. #7|*]

1869. ?? Are there mechanical/other ANALOGUES of 'ECONOMY'...
A/S   [Refs to be found ...]

1870. Formal descriptions of PERFECT COMPETITION #31
S/A   [Princeton #246*]

1871. IDENTIFICATION of models with NOISE #33
S/A   [Stoney Brook #210*]

1872. CONVEX PROGRAMMING/global convergence   #111
S/A   [SOL #79-1* (recursive substitution)]

1873. Dispersed information/uncertainty/approx.rational equilib. #19
S/A/A [U.Penn. #79-28*]

1874. INDEX NUMBERS:exact results/applications   #101
S/A   [Stanford #250*]

1875. DISTANCE between prob.distrns.VIA 'location estimation' #102
S/A   [UCB #554; cf.,ROUSSAS,Contiguity (CUP)]*

1876. Optimal equilib. for repeated games/imperfect monitoring #192
A/A   [RADNER,R.,Paper?B*]

1877. Measures of RISK-AVERSION   #34
S/A   [Harvard #366*]

1878. Number of 'blocking coalitions in CORE-equivalence thm
S/A   [UCB #240*]

1879. Stochastic evolution of self-fulfilling expectnl equilib. #152
S/A/A [Cornell #201*]

1880. Optimal depletion of exhaustable resources   #97
A/A   [Cornell #203*]

1881. Unit-cube alg.(lattice)criterion for block preferences #14
S/A   [N-W U. #372*]

1882. Mathematical structure of PREFERENCE-DUALITY theory #167
S/A   [U.Toronto #7901*]

1883. Existence of infinite-horizon competitive programs #168
S/A   [Cornell #202*]

1884. Interpolation and smoothing on spheres(SPLINES) #169
S/A   [U.Wisconsin #584*]

1885. Globally convergent condensation(GEOM.Programming) #170
S/A   [Stanford 79-11*]

1886. Asymptotic CORES/balance for large replica games #171
A/A   [Stoney Brook #215*]

1887. Iterative allocation with bidding #9
S/A   [UCB #20* (pp148)]

1888. Persistent disequilibrium
A/A   [UCB ?DG*]

1889. Constrained optimization over sequence-spaces #151
S/A/@ [USC #7826*]

1890. Probability that a 'random graph' is CONNECTED
S/A/@ [UCB ORC 79-5*]

1891. Uniqueness theorems for COMPETITIVE ALLOCATION #48
S/A [Minnesota 79-108*]

1892. Definitions/comparisons of EXPECTED INFORMATION #52
S/A [Princeton #239*]

1893. Evolution of (truncated) NONLINEAR FILTERS
S/A/@ [Princeton ## 240,242]*

1894. Utility- representations for partial orders #150
S/A/@ [UCB #IP-264*]

1895* Classification of DISEQUILIBRIUM REGIMES #149
S/A [UCB #IP-271*]

1896. Convergence to the CORE: generic examples #154
S/A/A [USC #7935*]

1897. Vector-valued LIAPUNOV functions for set-valued flows #1
S/A/@ [USC ##7702*]

1898. Representations of INFORMATION STRUCTURES #83
S/A [Stanford #271*]

1899. Optimal control over BSs #206
S/A/@ [USC #7706*]

1900. N-person dynamic games: trajectories/equlibria #158
S/A/A [USC #7711*]

1901. Manipulatory games ...   #143
S/A [U.Minnesota #79-111*]

1902. Dynamic oligopoly   #157
S/A [U C San Diego #79-23*]

(Quasi-)METRICAL TOPICS FROM RAAG MEMOIRS:VOLS I-IV (Inst.)

1903. Geometric/alg.N-W/(alg.)vector fields:Interrelations
S/@/A [R1:1-54;55-65(Table T)]

1904. Incidence matrice/topological interpretations of N-Ws
S/A [R1,68-111]

1905. Co-ord.-free topological calculations for N-W
S/A [R1,113-171]

1906. DUALITY of MESH and JUNCTION N-W representations
S/A [R1,172-178]

1907. DECOMPOSITION of admittances/impedances
S/A/@ [R2,118-130; #1885 (above)]

1908. NETWORKS as 2-D complexes containing surfaces
S/@/A [R2,84-117]

1909. Topological procedures for network SYNTHESIS
S/A/A [R2,32-41]

1910. Topological 'formulae' for 4-terminal N-W
S/@/A [R2,5-31]

1911. Trees/co-trees/multi-trees/multi-co-trees for N-W
δ/@/A [R2,42-83]

{ More examples from RAAG I-IV appear later in the list }

1912. SYMBOLIC vector/dyadic analysis
δ [SIAM J.Comp. 8(1979)306-319?*]

1913. Boundedness/stability for nonlinear ODEs
δ/A [MR 22A #9673 (long rev. of CESARI,L.(Ergib.)]*

1914. Rates of convergence in the Central Limit Theorem
δ/A [Pitman Res N #62(pp251,1982;HAAL,P.)]

1915. Systematic handling of FUNCTIONAL OPERATORS
δ/@ [JACM 8,168-185*]

1916. { Geometry of Spatial Forms }
δ [GASSON,C.(Horwood,1983;pp561);elem.+]

1917. Logics of scientific knowledge(analogy ...) 'Philos'!
A/δ [TAVANEC,P.V.,ed.(Reidel,1970;pp429)]

1918. { Fixed-point algorithms and applications
δ/A { [KARAMARDIAN,S.,ed.,bk.(1977;pp494)]

1919. { Matrix-DERIVATIVES }
δ/@ [ROGERS,G.S.(Springer,1980;pp209)*]

1920. { Computable Aanalysis }
S/A [ABERTH,O.,bk.(1980;pp187)]

1921. { Theorems/Problems in FUNCTIONAL ANALYSIS
S/@   [KIRRILOV,A.A.,et al.(Springer ?LNM;pp350]

1922. { Counter-Examples in TVSs
S/@   [LNM #936 (pp179,1982)]

1923. Inequalities and Uncertainty Principles
A/A [J M Phys 19(1978)461-466?*]

1924. Formal calculation of OPTICAL ABERRATION COEFFICIENTS
S/A [BUCHDAHL,?H.(Dover)* (?SAM)]

1925. Automated (nonlinear) CIRCUIT DESIGN
S/@/A [IEEE Proc.70(10)(10/82)1210-1228 (?SKELBOE,S.)]

1926. COMPOSITION-BASES for representation of classes of fns
S/@ [AMM 87(8)(1980)?*]

1927. Definitions of METRICS from MIN. ARCS on closed curves
S/A [SEE: Russian works on planar BVPs ...]

1928. For any rectifiable curve in C: mod(arg z) = O(1/mod(z))
S/A [SEYFULLAEV (Baku U.);Paper on BVPs]

1929. METRICS on configuration spaces in stat.mech.
δ/A/A [A-W Encycl.Math. Vol.5 (RUELLE,D.) ...]*

1930. DICTIONARY OF SYMBOLS in math.logic
δ [FEYS/FITCH,bk. (Inst.)]

1931. Applications of NEVANLINNA THEORY to MATH. LOGIC
δ/A/A [RUBEL,L.A.(U.Illinois,Urbana);Lec.,IHES]

1932. RECOVERY of mass distributions from SPECTRAL DENSITY fns
δ/A/A [Adv.Math.Suppl.Studies #3(1978)45-90?*;MR 80i #34029]

1933. Common fps of self-maps of P-O complete MSs
δ/A/A [PAMS 77,365-368;MR 80i #54053]

1934. Anal. capacity/approx^n (Russ M.S./Vitushkin, AG)
δ/A

1935. Non archimedean metric/top^l (J de Groot)
δ/A

1936. Approx^n of Manifolds by real alg. sets (Russ MS 37 (1982) 1-59 N.V. Ivanov)
δ

1937. Bounds for 'Stationary Phase' (SIAM JMA n5 (1974) 19-29 F.W.J. Olver)
δ

1938. Ext^n of 'Stationary Phase' Method (SIAM JAP M 15 (1967) 915-923 D. Ludwig)
@/δ

1939. Global construc^{tn} of Fractals (PRSL A399 (1985) 243-275 MF Barnsley.+)
δ

1940. Lusternik/Schnirelmann CATEGORY (Survey)
S   [Topology 17(1978)331-348?*]

1941. Collection of CONTRACTIVE DEFINITIONS
S   [Math.Seminar N Kobe U.(2)(1978)229-235?*]

1942. Random Differential Inequalities
S/A/@   [LADDE/LAKSHMIKANTHAM (?AP),Proof copy*]

1943. LYaPAS:programming language for logic/coding
S/A   [GAVRILOV,M.A.,et al.,eds.(?ACM)*]

1944. Numerically IMPLEMENTABLE topological procedures
S/@   [TAMS 1979+ ... ?]

1945. Inequality-form of Cauchy integral thm(n-conn. domains)
S/@   [MR 80m #30046]

1946. SIMPLICIAL SETS/foundations of analysis
S/@   [LNM #753,113-124?*]

1947. Topologies preserving algebraic structures
S/A/@   [Ann.U. Ferrara Sez VII(NS) 23(77) (1978),79-95]

1948. Elements/Exercises in Stochastic Analysis
S/@   [Kybernetika(Prague)Suppl. 14(1978):#1-3,p55-;4-5,p36-]

1949. LOCALIZATION in rings/algebraic defns.of NEIGHBOURHOOD
δ/A/@ [SEE,e.g.,HARTSHORNE,R.,Algebraic Geometry (SCHEMES)]

1950. Applications of forms of REGULARIZATION in elasicity
δ/A [PARTON/PERLIN:Int.Eqs in Elast.(MIR)*]

1951. CONTOURS/HOMOTOPY/ ... on VARIETIES
δ/@ [MR 25 ##5157/8/9;5162;J M Mech 11(1962)615-646;
       Riv.M U.Parma 10(1959:171-182;257-263]

1952. REDUCTION of 2-D potl.problems to Abel integral eqs.
δ/A [ZAMP 14(1963)675-681;MR 28 #4139]

1953. CONSTRUCTION of analytic functions without radial limits
δ/@ [PAMS 15,335-336]

1954. CONSTRUCTION of entire fns with spec. asymptotic behaviour
δ/A [EVGRAVOV,bk (G&B)]

1955. DEGREE of a rational matrix-function(N-W Theory)
δ/A [J SIAM 11(1963)645-658;MR 28 #822]

1956. NORMAL metrics (d(A,B)>0 if A,B,are disjoint sets)
δ/A [AMM 72998-1001?*]

1957. APPROXIMATION procedures in CONFORMAL MAPPING
δ/A [FIL'CAKOV,P.F.,bk.(pp531;Russian);MR 32 #1337]

1958. DISTANCE-SETS for the Cantor set
A/δ [AMM 72,725-729]

1959. { BOUNDS are related to LIMITATIONS ... }
δ/A/A [e.g.,in DAC,Coding Theory,...]

1960. Simultaneous operational calculus
δ/@ [DITKIN/PRUDNIKOV,bk*]

1961. Practical applications of CONVEXITY
δ/A [SIAM Rev.19(1977)221-240?*;QM,colour vision,decisions,..]

1962. Metrization of hyperspaces of MSs
δ/A/@ [Fund.Mat.94,191-207;MR 55 #6373]

1963. NEARNESS and metrization ...
δ/A/@ [LNM #540,518-547;MR 56 #9496]

1964. COMPLETENSS of USs and PSs
δ/A/@ [Colloq Mat 38(1977)55-62;MR 56 #16574]

1965. Jordan PAIRS (generalization of Jordan algebras)
δ/@ [LNM #460;MR 56 #3071]

1966. RELAY automatic-control systems
δ/A   TSYPKIN , bk.*;MR56 #4980]

1967. Operational methods for PDEs
δ/@ [MASLOV,bk.*;MR56 #3647(=Ztrlblt)]

1968. Topologies in sequence spaces VIA TOEPLITZ MATRICES
δ/A/@ [Michigan M J 5(1958)139-148?*;MR 21 #812 (2 cols.)]

1969. SET- and FUNCTION- distances
δ/A/@ [Colloq M 6(1958)319-327;MR 21 #2721]

1970. Fundamental theorems on ASYMPTOTIC SERIES
δ/A/@ [J Anal.Math.:5(1956/57)315-320;4,341-418;
        MR 18 #890;MR 211479]?*

1971. ULTRA-hyperbolic distance
δ/A/@ [e.g.,JONES,D.S.,Generalized Functions (CUP,1982)310-337]

1972. BOUNDS for the Banach-Mazur distance
δ/A [MR 81b #46038; ?SINGER,Bases in BSs]

1973. {Combinatorial complexes(A theory of algorithms)
δ/@   [SELLERS,P.H.,bk(pp184;?Birkhauser)]

1974. {Combinatorial problems/examples
δ   [LOVASZ,L,bk(?N-H)]

1975. Problem-analysis(various classes)
δ/@/A [Proc.(AP,1977;pp514);MR 80h #0004]

1976. Computable probability SPACES
δ/A [Act Math 103(196089-122(RANKIN,B.)*]

1977. Suprema of quotients of matrix norms
A/@ [Numer.Math.4(1962)114-116;MR 26 #889]

1978. Conformal METRICS on Riemann surfaces
δ/A/@ [Nagoya M J 21,1-60;MR 26 #1451]

1979. HIDDEN VARIABLES:various arguments
δ/A [Amer.J Phys. 29(1961)478-484]

1980. APPROXIMATELY CONTINUOUS transformations on compact MSs
δ/A [BAMS 68,488-493;MR 25 #5499]

1981. Partial sums of zeros of polynomials
δ/A [QJM(Ox) 13,151-155]

1982. Hilbert spaces NONassociative scalars
δ/@ [Math.Ann. 154(1964)1-27]

1983. Invariant manifolds(?Dynamical systems theory)
δ/A [LNM #583;MR58 #18595(long)]

1984. {General Topology/Outline of General Topology
δ/A/@ [ENGELKING,R. ,bks.(PWN(1977)/N-H(1965))]

1985. {HYPERSPACES of sets
δ/@ [NADLER,S.B.,Jr(Dekker,1978;pp707)]

1986. A CV-Rolle's theorem
δ/@ [AMM 74,452-453*]

1987. Topological extensions of COMPACTNESS
δ/A/@ [Tract,Math.Centrum A'dam(1968);van der SLOT]

1988. Constructive Real Analysis
δ/A   [GOLDSTEIN,A.A.(H&Row,1967;pp167   ]

1989. Dictionary of mathematical/physics SYMBOLS
δ/A   [POLON,D.D.,ed.(Odyssey,NY,1965,pp333+)   ]

1990. Roumanian/English mathematical DICTIONARY
δ   [GOULD,S.H./OBREANU,eds.(AMS,1967;pp51)   ]

1991. Probability measures on MSs
δ/A/@ [PATHASARATHY,K.(AP,1967;pp276)   ]

1992. BOUNDS on functions of MATRICES
δ/A   [AMM 74,920-9268* (?ROSENBLOOM,P.C.)   ]

1993. B-W-type theorems VIA LIGHT/HEAVY sets
δ/@   [AMM 72,1007-1012*   ]

1994. Comparison of SEPARATION SPACES and PSs
δ/A/@ [AMM 71,158-161*   ]

1995. General topology:unusual topics ...
δ/A/@ [MAMUZIC,Z.P.(Noordhoff,1963)*   ]

1996. Logico-mathematical remarks on GAMES
δ/A   [STEINHAUS,?H.,AMM 72,457-468*   ]

1997. 'Compact City' (Urban design)
δ/A [DANTZIG/SAATY(Freeman,1973;pp224)]

1998. SOLITONS: theory/applications
δ/A [Riv.Nuovo Cim.(2)7(1977)#4,429-469]?*

1999. Spline Analysis (basic ideas;mainly 1-D)
δ/A [SCHULTZ,M.(P-H,1973;pp156)]

2000. Integral Geometry(various inverse problems)
δ/@ [SANTALO,?L.(A-W/CUP Encycl.M Vol 1;MR 55 #6340]*

2001. Boundedness in convergence-Vector Spaces
δ/A/@ [Canad.M Bull.18(1975)499-502]

2002. ESTIMATES of 3-D volumes of parallel X-sectional AREAS
A/A [SIAM Rev.6,131-133;MR 30 #509]

2003. Geometric Measure Theory
δ/A/@ [FEDERER,H.(Springer,1969;pp676) (Inst.);MR41#1976(long)]

2004. Solution of Ill-Posed Problems
δ/A/A [TIKHONOV/ARSENIN (Winston/Wiley);MR 56 #13604]

2005. Topology and FORCING (Set Theory)
δ/A/@ [Russ.M Surveys 1983,77-136*]

2006. Classical Harmonic Analysis
δ/@ [REITER,H.(OUP)*;MR 46 #5933(long)]

2007. 'Interactions' between CS and mathematics
δ/A [Proc.Sympos.Appl.M 15(AMS,1963)]

2008. Path-integrals in POLAR CO-ORDINATES
δ/A [Proc.RS A279(1964)229-235;MR:28 #158;41 #7378]

2009. ODE-OPERATOR analogues of Rouche's theorem
δ/@/A [MR 32 #6181;Dunford/Schwartz Vol 2]

2010. COMPOSITION-FACTORIZATION of entire functions
δ/@ [Indian J P&A M2(1971)561-571]

2011. CALCULABLE numbers ...
δ/A [MR:38 #5626;43 #5699]

2012. FORMALIZATION of the elementary Theory of Equations
δ/@ [Bull.Res.Council Israel 9F(1960)47-70 (ROBINSON,A.)]

2013. Proof that ALMOST-ALL subgroups of a Lie group are FREE
A/@ [J.Alg. 19,261-262;MR 43 #7491]

2014. RECONSTRUCTION of measures(restrictions to sigma-fields)
δ/A/@ [ROGERS,Hausdorff Measures (CUP,1971)*]

2015. APPROXIMATE TOTAL DIFFERENTIALS of maps:R(n)-->R(m)
A/@ [FEDERER,bk(Springer):Stepanov's theorem]

2016. Extn of Cayley-Hamilton thm to INFINITE block matrices
δ/@ [JLMS 135(1959)273-288*]

2017. Simultaneous DAC for functions of n complex variables
δ/@   [JLMS 135,264-272*]

2018. MODULE-theoretic solution of the MATRIX equation f(S)=T
δ/@   [JLMS 135,325-336*]

2019. Generalization of the HM/GM/AM ineq. to A-G means
δ/@   [JLMS 102,449-463*]

2020. Topological properties of {Morse fns} in Hom[R(n),R]
δ/@   [MILNOR J., The h-Cobordism Thm (bk)*]

2021. Integration in finite terms for LINE INTEGRALS
δ/@   [Comm.Alg.3(1975)781-795]

2022. Determination of EFFECTIVE SINGULARITIES in composed PS
δ/@   [JLMS 132,398-403*]

2023. NSS conditions for the representation of fns as LTs
δ/@   [JLMS 17388-92* ; KRASNOSEL'SKII, Orliczspaces(bk)*]

2024. Nonlinear DISTANCE CRITERIA in pattern recognition
δ/A/A [Information Sci 9,359-363; MR 55 #6993]

2025. PROBABILISTIC TSs
δ/A/@ [J M Anal Appl 34(1971)67-81 MR 45 #1208] / Schweizer/Sklar (bk)*

2026. BOUNDEDNESS in TSs
δ/A/@ [J M P et A 28(1949)287-320?*] ≡ { bdd sets in TS }
S-T HU

2027. 'Remainders' in extensions of TSs
δ/A/@ [MR 33 #4904]

2028. CF-expansions of bdd analytic fns/'topological analyticity'
δ/A/@ [Topol.Anal.(WHYBURN,G.;2nd ed.]

2029. Connectedness in SYNTOPOGENOUS SPACES
δ/@ [PAMS 15(1964)590-595]

2030. Homology and Feynman Integrals
δ/A [HWA/TEPLITZ,eds.,bk.*]

2031. Extension of the Muskhelishvili complex reprn to R(3)
δ/@/A [QAM 23,313-322 (PENROD ,D.);also other(ABC)refs]

2032. Abstract 'Gramm-Schmidt' orthogonalization procedures
δ/@ [HOFFMAN/KUNZE,Linear Algebra,p369]

2033. Elastostatic formulation of 'quasi-isometric deformation'
δ/A/A [Sem.,1962/63 1st Naz.Alta Mat.,Vol ii(1965)462-473]

2034. DISTANCE-MATRICES of graphs / bks on Graph-Th/NW
δ/A [QAM 22,305-317*]

2035. Differential topology in smooth BSs
δ/@ [Topics in Topology (BOLYAI)589-608*]

2036. Sets of topologies COMPATIBLE with some metric(s)
A/@ [Topics in Topology(BOLYAI)215-222*]

2037. ALMOST-INVERTIBILITY of suspension functors
A/@ [Topics in Topology(BOLYAI)97-112*]

2038. Metrizability/algebraic boundaries of compact convex sets
A/@ [Topics in Topology(BOLYAI)55-60*]

2039. Nearness and metrization
A/@ [LNM #540,518-548 (?HERRLICH)]

2040. Generalized intervals/topology of arbitrary posets
δ/A/@ [Czech M J 26(101)(1976)527-540;TAMS 187,103-125;
       MR 55 #12593]

2041. Distance-functions on ordered semigroups
A/@ [Monats.M83(1977)223-251]

2042. Distance-Functions
A/@ [Proc 4th Prague Sympos.(1976)71-80* (REICHEL,H-C)]

2043. SUMMABILITY (Survey/tauberian theorems)
δ/A/@ [LNM #107*]

2044. Generalizations of WARING'S PROBLEM/MUNTZ'S THEOREM
δ/@ [JLMS 137,98-116;Interpolation & Approxn.(DAVIS,P.J.)]

2045. Effective Algebraic Number Theory
δ/A/@

[STOLARSKY,K.(Dekker,1974;pp329)*]

2046. INDISTINGUISHABILITY (mod.'profile functions')
S/A   [SCHWEIZER/SKLAR,Statistical MSs,bk*]

2047. Approximate:LIMITS/DERIVATIVES/TANGENTS/NORMALS
S/A/@   [FEDERER,H.,Geom.Measure Th.(Springer) (Inst.)]

2048. m-Based 'Minkowski inequality' (m:= measure)
A/@   [FEDERER,bk,p277-]

2049. Classes of PSEUDOMETRICS in BSs
S/A/@   [SINGER,Bases in BSs,Vol.2,p422-]

2050. Metric properties of information-theoretic ENTROPY
S/A/A   [KOTZ,Recent Adv.Infn.Th. (bk)*]

2051. INDECOMPOSABLE positive-additive functionals
S/@   [JLMS 41(1966)318-322;PCPS 58,196-205]*

2052. ESSENTIAL VARIABLES for n-variable functions
S/@   [JLMS 41,333-335*]

2053. Network Synthesis (general treatment ...)
S/A   [TUTTLE,D.F.,Jr.(Wiley/C&H,1958;pp1175);MR 21 #6204]

2054. Etude des Sommes Exponential
S/A   [SCHWARTZ,L.bk*;MR21 #5116]

2055. DAC of meromorphic functions in VALUED FIELDS
δ/@ [ROBERTS J.B.,Pacific J M 9(1959)183-193;MR21 #4319]

2056. MULTIPLE Wiener-Ito integrals/sums of dependent RVs
δ/A/@ [LNM #849 (MAJOR,P.;pp127,1981)]

2057. Quantum logic/INDUSTRIAL SYSTEM CONSTRUCTION
δ/A [GALAMBOS,S.,Hungarian Inst.Buiding Sci.,1979]

2058. Information theory in PLANNING problems
δ/A [MULLER,F,Publ.Pollak Mihaly HS PECS,1975]

2059. Graphical representations of complete MSs
δ/A JLMS(2)2,727-735* (HOLMES,R.A.)]

2060. Distribution of EXTREMA of RVs / bk* GUMBEL
δ/A [THOMPSON,J.A.,bk*,Ch 10]

2061. STAGES of definition (rel.to frameworks)
δ/@ [KOCK,A.,Synthetic Diffl.Geom.(CUP)*;Ztrlblt 466/7]

2062. h-Diametral fp-theorems
δ/A/@ [TASKOVIC,M.R.;Publ.Inst.Math(NS)27(41)249-258(1980);
Zntrlblt 467:54037*]

2063. Fp-theorems for NONARCHIMEDEAN PROBABILISTIC MSs
δ/A/@ [Publ.Math.25(1978)29-34(ISTRATESCU);Bull.Mat Soc Sci.
Math. RSR(NS)24(1972)359-362(HADZIC);Zntrl 467:60062]

2064 (i) Analogue of ALGEBRAIC FUNCTIONS for INFINITE MATRICES
δ/A

[JLMS 34(1959)273-280*]

2064. (ii) Analogue of Cayley-Hamilton thm for INFINITE MATRICES
δ/A [JLMS 34,281-288 (IBRAHIM,S.A.)]

2065. Geometry of Moment Spaces
δ/A [KARLIN,S./SHAPLEY (AMS Memoirs #12)*]

2066. Summability in TOPOLOGICAL GROUPS
δ/A/@ [Math Z: 96,259-278;103,129-138(PRULLAGE,D.L.);
       MR:35 #3323;37 #6646 (!!!!)]

2067. INSTABILITY of the essential spectrum of OD-operators
δ/@ [JLMS 43(1968)647-654* (McLEOD,J.B.)]

2068. Propagation of RELATIONS
δ/@/A [AMM 75,649-652 (MOSS,R./ROBERTS]

2069. Estimate of Prob{ xy = yx :x,y in G }
A/@ [Pacific J M 82,237-247]

2070. Estimates of EIGENVALUES from data in INVARIANT SUBSPACES
δ/A/@ [FICHERA,G.(Pitman)*]

2071. Enumerative formulae for ALGEBRAIC CURVES
δ/A/@ [PCPS 54(1958)399-416 (MACDONALD,I.G.)]

2072. Some extended contraction principles
δ/A/@ [PCPS 60,439-447 (EDELSTEIN,M.)]

2073. Optimal ... GRAPH-CYCLES/function approximation
δ/A [Numer.Mat.39,65-84* (GOLITSCHEK)]

2074. INTRINSIC METRICS in a MS
A/@ [AMM 73,937-950 (GLUCK,H.)]

2075. Analogues/dissimilarities:fns of 1 CV/n CVs
δ/@/A [KRANTZ,S.G.(?Wiley)*]

2076. Fully-abstract FINITE APPROXIMATIONS TO INFINITE MODELS
δ/A/@ [CMU CS-84-134*]

2077. HOMOTOPY concepts for FORMAL LANGUAGES
δ/A [Stud.Appl.M 66(1982)171-179?*]

2078. ANISOTROPIC function spaces
δ/@ [Anal.Mathematica 10(1984)53-77-96(TRIEBEL);M Nachr,1976]

2079. Approximation of CONVEX fns by RATIONAL fns
δ/A [Anal.Mathematica 10,15-21 (HATAMOV,A.)]

2080. Bdd linear map X-->X with no proper closed subspace
δ/@ [Bull.LMS 16(1984)337-401 (READ,C.J.)*]

2081. Problems/Theorems in ANALYSIS
  δ   [GELBAUM(Springer,1982;pp228)]

2082. Theorems and Problems in FUNCTIONAL ANALYSIS
  δ

[KIRRILLOV et al.(Springer,1982;pp330)]

2083. INVERTIBILITY of matrices over weakly noncommut. rings
δ/@ [MR 53 #470]

2084. ANALOGUES of DETERMINANT
δ/@/A [BEREZIN,?F.A.,The Method of 2nd Quantization (?AP)* ;
LEITES,D.A.,Usp.Mat.Nauk 30(1975)156-183]

2085. Distribution of EIGENVALUES of n-sections of inf. matrices
δ/A/@ [Lin.Alg.Appl.13(1976)185-199 (LUDWIG,A.)]

2086. n-Increasing fns.($B^2 >= A^2 ==> f^2(B) >= f^2(A)$)
A/@ [Usp.Mat.Nauk 30(1975)235-236]

2087. Approx. of A(1)...A(n) by B (n large,A(j)>=0;B>0,matrices)
A [MPCPS 79(1976)521-530;HAJNAL,J.,MR 53 #490]

2088. RADON transform (general treatment/applications)
δ/A [DEANS,S.R.(Wiley,1983;pp289)*-]

2089. CALCULUS in locally compact spaces
δ/A/@ [KELLER,H.H.,LNM #417*;MR 55 #13466]

2090. 'Gibbs phenomena'for genrl.cgce.of genrl.Fourier series
A/@ [JLMS(1953)148-156*]

2091. Various topics in COMPLEX ANALYSIS(detailed)
δ/A

[SEGAL,S.L.(Elsevier/N-H,1981;pp715)*-]

2092. Saddle-point method for MULTIPLE INTEGRALS
A/@ [FEDORJUK,bk(Nauka,1977;pp277);MR 58 #22580]

2093. MULTIGRID METHODS (for PDEs, ...)
S/A/@ [LNM #960]?*

2094. INVERSE PROBLEMS in Nevanlinna theory
S/A/@ [Acta Math.138(1976)83-151*;MR 58 #28502]

2095. Evaluation of singular integrals('division of GFs')
S/A/@ [Bull.IMA 12,203-206 (ATIYAH,M.F.)]

2096. { Summability Through Functional Analysis }
S/A/@ [WILANSKY,A.(N-H,1984;pp318)*]

2097. { Handbook of SET-THEORETIC TOPOLOGY }
S/A/@ [KUNEN,K./VAUGHAN,eds.(N-H,1984;pp1260)*--]

2098. Open CV-problems in 'minimal settings'
S/A/@ [Linear and Complex Analysis Probl.Bk.(LNM #1043)]

2099. Fast 'interior- POLYTOPE algorithm'(Linear Programming)
S/A/A [KARMARKAR (Bell Labs.,1984/later papers*-)]

2100. APPLIED MATHEMATICS:Various techniques
S/A *[PEARSON,C.E.,ed.Handbook(vN-R);HINCHEY,F.(Wiley Eastern)]

2101. Eigenvalues of COMPOSITE MATRICES
δ/@ [PCPS 57,37-49]

2102. Analytical proofs of basic TOPOLOGICAL results
δ/@ [AMM 85,521-524*]

2103. COMBINATORIAL RESULTS(diverse collection with proofs)
δ/@ [LOVASZ,L.(N-H,1979;pp551)]

2104. Generalized VARIETIES as LIMITS   /PLMS (Wallace)
δ/A/@ [J M Mech.13(1964)673-692;MR 29 #4873]

2105. BOUNDS on zeros of polynomials(unified treatment)
δ/A [Thesis(Greek!);MR 29 #4875]

2106. Unified theory of TS,PS,and US(generalized topologies)
δ/A/@ [Dok.A.N.SSSR 156(1964)21-24;MR 29 #5221]

2107. DECOUPLING of pairs of PDEs
δ/A [QAM 27,87-104*]

2108. DISTANCE between POLYTOPES in R(n)
A/@ [QAM 26,207-212*]

2109. STABILITY of chemical reactions VIA GRAPH-THEORY
δ/A/A [JCP: 60,1481-1501(I,II);62,773-775;Proc.(Topol./GT in
       Chem.(Elsevier)*);MR 50 #15756/7/8]

2110. STRUCTURE of M^n (M,matrix) VIA GRAPH-THEORY
δ/A   [SIAM J Appl M:14,610-639;762-777;MR 37 #1264/5]

2111. Distance estimates and probability laws
A/@   [Publ.Inst.Stat.U.Paris 23(1978)37-88?*]

2112. { Information theory and chemical analysis
δ/A   [ECKSCHLAGER,K.,et al.?eds.(Wiley,1985;pp152]

2113. AI in Continuum Mechanics
δ/A   [WONG,A.K.C.,Proc.ASCE (12/70)1239-1265?*]

2114. n-Dimensional ANALOGUES:N-W/Vector Anal./Theor.Physics
δ/A   [Proc.Large Engrng.Systems 2*(SAVAGE,G.J.,ed.)357-362]

2115. Approxn.of general optimization by seqs.of GEOM.PROGR.
δ/A/A [Indiana U M J 22(1972/73)531-550 (DUFFIN,R.J.,et al.)]

2116. { Research topics in ANALYTIC NUMBER THEORY }
δ/A   [POSTNIKOV,A.G.(Nauka,1971);MR55 #7895]

2117. METRICS on spaces of STATL.-MECH. INTERACTIONS
δ/A/A [Reports Math.Phys.12(1977)19-25;MR 57 #11564]

2118. PROPAGATION of convergence
A/A   [e.g.,BURRIL/KNUDSEN,Real Variables*,p 280-]

2119. { Computers in Mathematical Research }
δ/A/A   [Dept M/CS/Stats U Illinois Chicago (1985/86)]

2120. DISTANCES between EXPERIMENTS
A/A   [MR 52 #1956 (?Le CAM,L.)]

2121. NEARNESS structures(Survey/motivation)
δ/A   [MR 50 #3193]

2122. CONsimilarity (of matrices)
δ/@   [HORN/JOHNSON,Matrix Analysis(CUP,1985)*]

2123. SAM and NONlinear control theory
δ/A/A   [LN Control Infn. Sci.#58(1984)?*;MARINO,et al.]

2124. { Structure des Systemes Dynamiques(symplectic geom. }
δ/A   [SOURIAN,J-M,bk(Dunod,1970);MR 41 #4866]

2125. BOOLEAN calculus of differences
δ/A   [LNCS #101*;?Modern Switching Theory(bk)*]

2126. Complexity-INTERPOLATION for computer languages
A/A   [LNCS #117?*,p 400-]

2127. INVARIANTS of sequence transformations
δ/A   [SIAM J.Math.Anal.6(1975)97-104 (LAWERIER)]

2128. Prime factorization of AUTOMATA
δ/A

[Math.Systems Theory 11(1978)239-257 (DORFLER,W.)]

2129. Random motion on an INFINITE ABELIAN GROUP
δ/@ [PCPS:47(1951)756-762;48,368-;MR:13 pp363,663(GOOD,I.J.)]

2130. Saddle-point methods for MULTINOMIAL DISTRIBUTIONS
A/@ [Ann Math Stat 28(1957)*,861-868;MR 20 #386]

2131. INequivalent modes of DECOMPOSITION for n-D systems
δ/@ [IEEE Trans.Circuits Systems CAS-26 105-11(1979)]?*

2132. INTRINSIC distances on SHELLS(Elasticity)
δ/A/A [MR 52 #9753]

2133. 'RADON transforms'for GRASSMANN MANIFOLDS/MATRIX SPACES
δ/@ [Soviet M Dokl.8(6)(1967)*,1504-1507]

2134. Integral trans./Variational techs./DEs for DISTRIBUTIONS
δ/A/@ [TEODORESCU,P.P./KECS,W.(Abacus P)*;ZEMANIAN(Dover)*]

2135. Eigenvalues of FIELD-COMBINATIONS of matrices
δ/A [MR 50 #4629(with many REFS)]

2136. SAMPLING as the INVERSE of INTERPOLATION
A/A [e.g.,BROWN,B.M.,Math.Th.Lin.Systems*]

2137. REPRESENTATION of the SHIFT operator VIA exp{hD}
δ/@ [e.g.,BROWN,B.M.,Math.Th.Lin.Systems*]

2138. Galois-UNSOLVABILITY of basic sextic eq.in aniso.elast.
δ/A [J.Elasticity:9,1-20;321-325 (HEAD,A.K.)]*

2139. NETWORK SYNTHESIS as a RECONSTRUCTION problem
δ/A/@ [SEE modern bks on N-W Theory/MRI Symposia*]

2140. CONVOLUTION ALGEBRAS/inductively coupled CIRCUITS
δ/A [MR 56 #2678(KECS,W.)]

2141. { SYNERGETICS }
δ/A [Bk.(HAKEN,?H.;Springer);Revs Modern Phys 47,67-121;
       MR 57 #8813]

2142. SIMPLICIAL APPROXIMATION for DESIGN under uncertainty
δ/A/A [IEEE Trans Circuits Systems(IEEE TCS) CAS-24,363-72]

2143. The MEASURE OF A MATRIX(in circuit design)
δ/A/A [IEEE Trans Circuits Systems CT-19(1972)480-486]

2144. Vector Boolean algebra/calculus
δ/A/@ [IEEE Trans Computers C-25(1976)865-874]

2145. ANALOGUE-COMPUTER solns of systems of lin.INEQUALITIES
A/A [Automation and Remote Control 1969(3)435-455?*]

2146. POSITIVITY-preserving (linear) transformations
δ/A/@ [IEEE TCS:CT-19(1972)460-465;CT-12(1965)607-608]

2147. CONSTRUCTIVE matrix criteria for STABILITY
δ/A [IEEE TCS CAS-26(4/79)224-234]*

2148. A 'GENERALIZED-FUNCTION dynamical system theory'
δ/@/A [Proc NRL-MRC Conf.(Naval Res.Lab,Washington(AP, 1972)
pp 151-164 (KALMAN)]

2149. Divided differences in FINITE FIELDS
A/@ [IEEE Trans Comp.C-27(1978)232-238]

2150. Operator-approximations to the IDENTITY operator
A/@ [JLMS 42(1967)477-483 (OKIKIOLU);also:bks on Banach alg.]

2151. ITERATION-formulae for SINGULAR INTEGRALS
δ/@ [POGORZEL'SKII,Integral Eqs.(Pergamon)*,p 445-]

2152. A METRIC for numerical approximation over curves
A/@ [SIAM J Numer.Anal.21(1984)202-215?*]

2153. { Green's Functions and Transfer Functions Handbook }
δ/A [BUTKOVSKIY (Ellis Horwood,1982+) (RMCS)]

2154. Papers on THEOREM-PROVING by computer
δ   [SIEKMANN,ed.;2 vols;Springer]
   /WRIGHTSON

2155. { Qualitative Methods in Elasticity }
δ/A/A [VILLAGGIO,P.(Sithoff,1977)*-]

2156. Extended B-W Theorems(VIA 'light/heavy maps'
δ/@ [AMM 72,1007-1012(KIMBER,J.E.)]

2157. Matrix-analytic function theory/sets of (O)DEs
δ/A/@ [LAPPO-DANILEVSKY (Chelsea,1953) (?Inst.)]

2158. NONarchimedean fields/ASYMPTOTIC EXPANSIONS
δ/A/@ [BAMS 83,231-235 (REV of LIGHTSTONE/ABR.ROBINSON,bk)]

2159. 'Uncertainty Principles' for RECONSTRUCTION problems
δ/A/A [Duke M J 42(1975)661-706(LOGAN,B.F.);MR 54 #3264]

2160. Essential undecidability in terms of NATURAL BOUNDARIES
δ/A [SEE:various expositions in Foundations bks ...]

2161. Differentiation of multiple integrals over VARIABLE DOMAINS
δ/A/@ [BAMS 75,172-173;HUNT,J.N.]

2162. (i) CLOSE-TO-CONTINUOUS functions
δ/@ [BAMS 74,1036(Abstract);IRUDAYANATHAN,A.]
δ/A (ii)'New proofs' that a continuous fn ATTAINS its bounds
  [BAMS 74,1082;HARDY,'PM',Sec.105]

2163. { Mathematisches Worterbuch }
δ [NAAS/SCHMID,eds.;TEUBNER/PERGAMON,1961)   (Inst.)]

2164. { Algebraic Methods in Semantics
δ/A [NIVAT,et al.,eds.(CUP,1985)*]

2165. Systemes de Polynomes(analogues of Taylor series ...)
δ/A/@ [Papers in P&A M #35(Queen's U,Ontario,1973);ROBERT,A.]

2166. Economy and power of NOTATION(Subharmonic fns)
δ/@

[e.g.,MAURIN,K.,Analysis,Vol 2*378-384]

2167. Heat-functions:developments analogous to analytic fns
δ/A/A [WIDDER,D.,The Heat Eq.(bk);1979+]

2168. CV-analogues of Rolle's Theorem
δ/A [AMM 74,452-453]

2169. Operational calculus in COMMUTATIVE ALGEBRAS
δ/@ [J M Pures Appl.(9)B3(1954)147-186;WAELBROECK,L.A.]

2170. STOCHASTIC parallel displacement
δ/@/A [LNM #451*(ITO,K.)]

2171. Simple approaches to SPECTRAL SEQUENCES
δ/@ [BAMS 76,599-605 (MITCHELL,B.)]

2172. 'Homotopy Theorems' in function theory
δ/@ [AMM 76,778-787;REDHEFFER,R.]

2173. Intersection/covering properties of CONVEX SETS
δ/A [AMM 76,753-766;CHAKERIAN]

2174. { BIBLIOGRAPHY on Schlicht Functions }
δ [BERNADI;pp157;MR 34 #2849;NR041-019 IMM-351
   ONR(Courant Inst.)Report]

2175. Singularities of differentiable maps(Dissections(?SAM))
δ/@ [ARNOL'D,V.I.,et al.(Birkhauser,1985;2 vols) (Inst.)]

2176. { Mathematical Economics (intermediate level) }
δ/A   [TAKAYAMA;2nd ed.,CUP,1985;ppc730] ✳

2177. n-D conformal maps/plane potential theory/Bieberbach
       Conjecture as an 'all-or-nothing result'
δ/@   [HENRICI,P.,Appl/Compl CV III]*-

2178. Functional-analytic proofs of APPROXIMATION THEOREMS
δ/@/A [BAMS 76,483-489 (RUBEL/TAYLOR)]

2179. { Applic.of generalized continued fractions to APPROXIMATION
A/@   [KHOVANSKII (Noordhoff,1963;pp209)]

2180. INTRINSIC METRICS
A/@   [AMM 73,937-950*]

2181. NSS conditions for MULTIPLICATIVE DECOMPOSABILITY
δ/A   [AMM 75,350-357 (DOTSON,W.G.)]

2182. ZERO-distributions of f'(for ENTIRE functions f)
δ   [BAMS(?AMM) 75(10/68)829-839 (MARDEN,M.);AMS Colloq.
        Publ.(SCHAEFFER/SPENCER)*,Appendix]

2183. Approximation of ASYMPTOTIC MISS-RATIOS for data-bases
A/A   [J Comp.Syst.Sci.14(1977)222-250 (FAGIN,R.);MR 58 #25069]

2184. { McG-H Dictionary of Physics and Maths. }
δ   [LAPEDES,D.N.,ed.;pp1074+ (nontrivial);MR 58 #8857]

2185. COMPLEXITY of trajectories in dynamical systems
δ/A   [Usp.Mat.Nauk 33(1978);BRUDNO,A.A.]

2186. Artin's Conjecture/Euclidean algorithm in alg.no.fields
δ/A/@ [Invent.Mat.42(1977)201-224;MR 58 #576(LENSTRA,H.N.,Jr)]

2187. An Unfinished Dialogue With G.I.Taylor
δ   [J Fluid Mech.70(1975);BATCHELOR,G.K.]

2188. Repeated integrals expressed as single integrals
δ/@   [J Approx. Th. 5(1972)276-307 (DAVIS,P.J.)]

2189. 'Almost-all' for ALGEBRAIC VARIETIES
A/@   [LNM #997*,p158-   ]

2190. Upper bounds for ALGEBRAIC K-THEORY
δ/A/@ [LNM #1046*,329-348(SNAITH,V.)   ]

2191. CONSTRUCTIVE category theory
δ/@   [LNCS #118*,563-   (KANDA,A.)]

2192. Metric spaces in LINGUISTICS
δ/A   [LNCS #53*537-542 (VIANU)]

2193. Martingales in Probability and Analysis
δ/@   [DURRETT (Markham,Chicago,1984)*]

2194. Surveys of GEOMETRY(modern approaches ...)
δ/@   EVES,H.,2 vols.(Allyn&Bacon,c1965);BERGER,M.(Springer)} ✗—

2195. Recurrent INEQUALITIES
δ/A [PLMS(3)17,683-699 (REDHEFFER,R.)]

2196. Generalized CONVEXITY
δ/@ [PLMS(3)17644-652 (DAVIES,E.B.)]

2197. Set-theoretic generalizations of DAC
δ/@ [PLMS(3)1152-162 (MAJSTRENKO)]

2198. ODE-operator ANALOGUE of Rouche's theorem
δ/A [MR 32 #6181;DUNFORD/SCHWARTZ II]

2199. Approximately-total DIFFERENTIALS of maps
δ/A/@ [FEDERER,H.,Geometric Measure Theory(Stepanov's thm)]

2200. Handbook of SET-THEORETIC TOPOLOGY
δ [KUNEN,K./VAUGHAN,eds.,pp1273;N-H,1984]*--

2201. FORMALIZATION of the elementary Theory of Equations
δ [Bull.Res.Council Israel 9F 47-70 (1960) (Abr.ROBINSON)]

2202. PROOF that ALMOST-ALL subgroups of a Lie group are FREE
δ [J Alg.19(1971)261-262;MR 43 #7491]

2203. GFs AS equivalence classes of approximating functions
δ/A/@ [Stud.Math.23(1963)119-140]

2204. Module-theory solution of polynomial MATRIX EQUATIONS
δ/@ [?JLMS #135,325-336]

2205. CONSTRUCTIVE proof of the MUNTZE Approximation Theorem
δ/A [AMM 75,342-350]

2206. { Foundations of Analysis in the Complex Domain }
δ [MR 36 #5309 (?CERNY,et al.)]

2207. Fixed-points of ITERATED MAPS without CONTINUITY
δ/A/@ [AMM 75,399-400]

2208. { The Hyper circle in Mathematical Physics }
δ/A/A [SYNGE,J.L.(CUP,1956)*;MR20 #4073]

2209. Topological properties of {Morse fns} in Hom[R(n),R]
δ/A/@ [MILNOR,J.,Lecs.on the h-Cobordism Thm (Princeton)*]

2210. CV claculations in LISP(REDUCE)
δ [BIT 16(1976)241-256 (CAMPBELL,J.)]

2211. { Approximation procedures for CONFORMAL MAPPING
δ/A [FILCAKOV (bk)pp531,1964;GAIER,D.(Springer)]

2212. Generalizations of 'Mean Inequalities'
A/@ [JLMS #120(1955)449-463*]

2213. 'Cayley-Hamilton Theorem' for INFINITE BLOCK MATRICES
δ/@ [JLMS #135281-288]

2214. BOUNDS for classes of INFINITE MATRICES
δ/A/@ [JLMS #135,273-280]

2215. {RECONSTRUCTION of MEASURES from various 'restrictions'
A/@ [ROGERS,C.A.,Hausdorff Measures (CUP)*]

2216. 'Calculable Numbers'
δ [JACM 15,275-299*;MR:38 #5626;43 #5699]

2217. Composition-factorization of entire functions
δ/@ [Indian J P&A M 2(1971)561-571]

2218. Interaction of MATHEMATICS and CS
δ/A [Proc.Sympos.Appl.M 15(?AMS,1963)]

2219. Path-integrals in polar co-ordinates
δ/@ [PRS A279(1964)229-235]

2220. REVIEW of Harmonic Analysis (REITER,H.) *
δ [MR 46 #5933 (long)]

2221. Topology and Forcing
δ/A [Russian Math.Surveys 38(1983)77-136*]

2222. Theory of Mathematical Machines
δ/A [BAZILEVSKII,Y.Y.,ed.(Pergamon P,1963;pp264)*]

2223. SMALL OBJECTS in Category Theory (?Defn of LARGE)
δ/A/@ [e.g.,FAITH,C.,Algebra,Vol.1(Springer)*,p213-]

2224. METRIZATION of the SPACE OF PLASTIC COLLISIONS
δ/A/A [KECS/TEODORESCU(Abacus P)*,p324-]

2225. { Fondements du Calcul Differentiel }
δ/A [Ver EECKE,P.(P.U.de France;2 vols;1983);anno.biblio*]

2226. { Singularities of Differentiable Maps(CONSTRUCTIVE) }
δ/@ [Birkhauser (Inst.)]

2227. A 'Gerschgorin Theorem' for TENSORS
δ/A/@ [Int.J.Control (3/86);BANKS,S.]

2228. Spaces of COUNTABLE TYPE(admit 'few distinct mappings')
δ/@ [Van MILL,J. (?Math.Centrum Tract) (?Inst.)]

2229. AI in CONTINUUM MECHANICS
δ/A [WONG,A.K.C./BUGHERELLO;Proc.ASCE(12/70)1239-1265]*

2230. Computation in ALGEBRAIC GEOMETRY (?SAM)
δ/A [AMM 89(1982)34-56 (WAVRIK,J.)*]

2231. Matrices with prescribed BOUNDS on EIGENVALUES/ELEMENTS
δ/A [Math.Computer Performance,LAZEOLLA,G.,et al.,eds.]*

2232. Riemann Mapping thm/Dgt FS/Reciprocal FS Thm VIA Fnl Anal.
δ/@ [WILANSKY,A.,Functional Analysis(Blaisdell)*]

2233. Extensions of PAPPUS' Theorem
δ/@ [GOODMAN,A.W.,AMM76,355-366]

2234. Algebraic reprn thms for ODEs/difference equations
∫/@ [AMM 76,366-373 (KLARNER,D.A.)]

2235. Reprn for ITERATED- POWER(a(1)...a(N):SUM{a(i)} fixed)
∫   [AMM 76,830-]

2236. Infinite LEGENDRE TRANSFORM/INVERSE
∫/@ [PCPS 57,547-560 (CLEMMOW,P.C.)]

2237. 'Krein-Milman Theorem' for POSETS
∫/@ [AMM 76,282-283 (BAKER,K.A.);BIRKHOFF,G.,Lattice Theory]

2238. Mean of x(t) relative to w(s,t)
∫/@ [AMM 76,252-261 (CASHWELL/EVERETT)]

2239. Topological/algebraic definitions of 1-CONNECTEDNESS
∫   [AMM 74,117-120;CHEVALLEY,C.,Lie Groups(for b/g)]

2240. Composition and inversion for ASYMPTOTIC FUNCTIONS
∫/A/@ [AMM 74,1095-1097]

2241. COMBINATORIAL TESTS of matrix orthogonality
∫/@ [AMM 74,1083-1084]

2242. General solutions of arbitrary BOOLEAN EQUATIONS
∫/@ [AMM 74,1074-1077]

2243. The Commutative Law of CREEP
S/A [ODQUVIST,?F.,bk (OUP,1st ed.,p16)]*

2244. Determinantal reprns of truncated TAYLOR SERIES/remainders
S/A [AMM 88,528-]

2245. Basic theorems for AP-FUNCTIONS and their 'FS'-reprns
S/@ [AMM 88,515-527 (COOKE,R.L.)]

2246. 'Subexact sequences'(DEVIATION from exactness ...)
S/@ [AMM 75,1087-1090]

2247. ANALYSIS OF MECHANISMS/MANIPULATORS (detailed formulae)
S/A [DUFFY,J.,bk (?Arnold,1980;pp419)* (?SAM)]

2248. Markov fields/log-linear CONTINGENCIES(Stat.Mech techn.)
S/A [Ann. Stats.8(1980)522-539]

2249. ZEROS of special sequences of polynomials
S   [LNM #1002*;M USSR Isvest. 9(1975)63-112(DZRBASJAN]

2250. SET-THEORETIC GENERALIZATIONS OF 'DAC'
S/@ [PLMS(3)1,152-162;MAJSTRENKO,P.]

2251. SIF as an ANALOGUE of RESIDUE (LEFM/CV)
S/A [Treatment VIA SINGULARITIES of PDEs ...]

2252. HOMOTOPY in formal LANGUAGES
S/A [Studies Appl M 56,171-179(FARMER,F.D.);
       MJapon.:20,21-28;23,607-613;Discr.M11,23-27(HOMOLOGY)]

2253. Structural analysis/SYSTEMS THEORY(graph theory)
S/A [Devel.Th.Appl.Mech.Vol.2*(WU,T.S.)]

2254. SINGULAR PERTURBATION/boundary-layer theory/linear ODEs
S/A/A [GOLDSTEIN,S.,pp41-67 of Contr.to Mech.(ed.,ABIR)*]

2255. Normal modes in NONlinear systems
S/@ [QAM 24,177-193* (ROSENBERG,R.M.)]

2256. BOUNDS on condition numbers of optimally scaled MATRICES
S/A [JACM 21,514-524* (FENNER,T.L./LOIZOU)]

2257.'Critical value'criterion for change-of vars(Lebesgue int.)
S/@ [AMM 76,514-520 (SERRIN,et al.)]

2258. 'Topological' vs 'Computability'approach to UNSOLVABILITY
S/@ [JACM 20,399-408 (MILLER,W.)]

2259. CORRESPONDENCE:Turing Machines/Axiomatic Problems
S/A [JACM 17,241-259 (WILLIS,D.G.)]

2260. ANALYSIS in the COMPUTABLE NUMBER FIELD
S/A [JACM 15,275-299 (ABERTH,O.)]

2261. General Systems Theory(Mathematical Foundations)
S/@/A [MESAROVICH,M.D./TAKAHARA,Y.(AP,1975;pp268)]

2262. Analytical/geometric approaches to CATASTROPHE THEORY
S/A [AMM:81,211-260;85,765-801 (CALLAHAN,J.)]

2263. Sufficient conditions for ASSOCIATIVITY in algebras
δ   [AMM 78,1107- (COUGHLIN,et al.)]

2264. { Meyers Handbuch uber die Mathematik }
δ   [MESCHOWSKI,H.,ed.;Bibliographisches Inst.,1972;pp1234!!]

2265. Are ALMOST-COMMUTING MATRICES near COMMUTING MATRICES?
A/@ [AMM 76,925-926 (ROSENTHAL,P.)]

2266. Prime MODELS/ALMOST-decidability
δ/@ [J Symbolic Logic 51(1986)412-420(MILLAR)]

2267. The STRENGTH of Nonstandard Analysis
δ/@ [J Symbolic Logic 51,377-386 (HENSON,C.W./KEISLER,J.]

2268. Separation-of-variables in n-D Riemannian MANIFOLDS
δ/@ [J M Phys.27(1986)1721-1736?* (KALNINS,E./MILLER,W.]

2269. 'DISCRETE quantum mechanics'
δ/A [J M Phys 27,1782-1790 (GUDDER,S.)]

2270. DECOMPOSITION into ATOMIC and DIFFUSE functionals
δ/@/A [Duke M J 27(1960)597-606 (GORDON,H.)]

2271. Extensions of Holder's Inequality
A/@ [PLMS(3)11311-326 (KALMAN,J.A.)]

2272. Extension of RESULTANT to pairs of ANALYTIC functions
      [Applicable   Anal.7(1977/78)191-205 (GOHBERG,I.C.,et al.)]
δ/@

2273. Extension of KOROVKIN'S APPROXIMATION theorems ...
A/@ [Various papers*;?Pitman Res.Notes]

2274. PRODUCT INTEGRATION over (in)finite paths/contours
δ/@ [DOLLARD,J./FRIEDMAN (A-W/CUP,1979)]

2275. Effective solns.of eqs.in free groups VIA SURFACE TOPOLOGY
δ/@/A [Topology 20(1981)133-145;CULLER,M.*]

2276. Functional-analytic proof of the RIEMANN MAPPING THEOREM
δ/@ [WILANSKY,A.,Fnl. Anal.(Blaisdell,1964),p68- (LAX,P.D.)]

2277. Handbook of SET-THEORETIC TOPOLOGY
δ/A [KUNEN/VAUGHAN,eds.,N-H,1984;pp1273]

2278. Analytic functions of MATRICES
δ/@ [MPCPS 55(1959)51-61 (AFRIAT,S.N.)]

2279. Practical applications of EXTERIOR CALCULUS / Flanders (bk)*'
δ/A [EDELIN,D.G.B.,bk,Noordhoff,1980;ALSO:more recent bks]

2280. Proof of Inv.-Fn.Thm VIA INTEGRABILITY conditions
δ/@ [?REF(? Rev. in AMM (?Allerendorfer,C.))]

2281. VOLUMES of domains in ORTHOGONAL/UNITARY SPACES
δ   [Q J M 20#79(1949)146-154 (PONTING,F.,et al.)]

2282. Wiener-Hopf equations in GENERALIZED FUNCTIONS
δ/@ [Proc.Moscow M S (1976) (VOLEVIC,et al.);transl.]

2283. δ { (i)Theory of Distributions(FRIEDLANDER;CUP,1982)
     (ii)Complex Analysis(KODAIRA;CUP,1986)
     (iii)Spectra of PD Operators(SCHECHTER,N-H,1971)
     (iv)(Topics in)Matrix Analysis(2 vols;HORN/JOHSON;CUP)}

2284. DE-COUPLING of pairs of PDEs
δ/A [QAM 27,87-104 (CHIU,H.)]

2285. 'Stokes-flow formulation' of (NONSTATIC) ELASTICITY
δ/A [QAM 27,57-65 (KANWAL,R.)]

2286. NEUTRIX CALCULUS(Introduction)
δ/A/@ [J Analyse Math.7(1959/60)281-399 (van der CORPUT)?*]

2287. Radon-Nykodim Theorem AS a result in PROBABILITY THEORY
δ/@ [AMM 85 (1978) 155-165; S.M. SAMUELS]

2288. Composite-function LIMITS in TOPOLOGICAL SPACES
δ/A/@ [AMM 84,49-52 (SAMUELS,S.M.)]

2289. TSs of CRYSTAL(Bravais)LATTICES
δ/A/A [Trends in Appl.P M to Mech.#2* (ROGULA)]

2290. CONVERSE of Rouche's Theorem
δ [AMM 89,302-305 (CHALLENER,D.,et al.)]

2291. The BOUNDARY of a TS
δ/@ [AMM 89,307-309 (SCOTT,B.M.,et al.)]

2292. Differentiation of ASYMPTOTIC FORMULAE
    A/@   [AMM 88,526-527 (NEWMAN,D.J.)   ]

2293. Estimates for ILL-POSED PROBLEMS
    S/A/@ [Math.of Computation 28(1974)889-907(FRANKLIN,J.N.)]

2294. STRUCTURE/APPROXIMATION in Physical Theories
    S/A/A [Sympos.;HARTKAMPER,A.,ed.(Plenum P,1981;pp264)*-   ]

2295. Nonlinear dynamics and TURBULENCE
    S/A/A [BARENBLATT,G.I.,et al.,eds.(Pitman,1983)   ]

2296. Lie-group techns for GENERALIZED SEPARATION-OFVARIABLES
    S/@

2297. PROGRAMMING with SETS (SETL examples)
    S/@   [SCHWARTZ,J.T.,et al.(Springer,1987)   ]

2298. ACCUMULABILITY (a generalization of SUMMABILITY)
    S/A/@ [Duke M J 27(1960)555-560   ]

2299. Generalized POWER-SERIES INVERSIONS/REVERSIONS
    S/A/@ [SIAM J Numer. Anal.9(1972)241-247(ESTES,R.H.,et al.)*]

2300. Matrices of MONOTONE KIND
    S   [SIAM J Numer.Anal.10,618-622(WILLSON,A.N.)*   ]

2301. POSETS of MATRICES/convergence of ITERATION
    S/A   [SIAM J Numer.Anal.9,97-104(VANDERGRAFT)   ]

2302. ASYMPTOTIC ANALYSIS/unification of QUADRATURE RULES
δ/A [SIAM J Numer.Anal.9,pp... (DONALDSON,J.D.,et al.)?*]

2303. Evaluation of CAUCHY TRANSFORMS (CV Cauchy integrals)
δ/A [SIAM J Numer.Anal.9,284-299* (ATKINSON,K.)]

2304. LISP (general presentation/examples)
δ [WINSTON,P.H./HORN,B.... (A-W,1981;pp430)*]

2305. GEOMETRY (modern approach/topics)
δ/@ [BERGER,Marcel;2 vols.(Springer,1985+)]

2306. LOGARITHMIC-MEAN inequalities
δ/A [AMM 81,879-883 (LIN,T.-P.)]

2307. Riemann surfaces and SERVICE CONNECTIONS(graph theory)
δ/A [SPILLERS,W.R.,ed.,Basic Questions of Design
Theory(N-H,1974)383-394 (GROSS,J.L.)]

2308. PARAMETRICES/BV SYMBOL CALCULUS
δ/@ [Math.Nachr.105(1982)45-149 (REMPEL/SCHULTZE)?*]

2309. BLOCK-GENERALIZED INVERSES(of matrices ...)
δ/@ [ARMA 61(1976)197-251?*(HARTWIG)]

2310. Power-series whose SECTIONS have ZEROS of LARGE MODULUS
δ/A [TAMS 117(1965)157-166]

2311. POSETS of (information) CHANNELS
δ/A/A [IEEE Trans.Infn.Th. I-T 13(1967)
360-365(HELGERT);MR 35 #6485]

2312. GENERALIZED Wiener-Hopf factorization
S/@ [SPECK,F.-O.(Pitman,1985)*]

2313. POSITIVE-DEFINITE FUNCTIONS and their generalizations
S/A/@ [Rocky Mountain J M 6(1976)409-434* (STEWART,J.)]

2314. FOURIER ANALYSIS (discursive treatment/diverse examples)
S/A [KORNER,T.W.(CUP,1988;pp588)]

2315. MSs of INTERACTIONS and the THERMODYNAMIC LIMIT
S/A/A [Reports M Phys.12(1977)19-25?* (MESSER,J.)]

2316. {'Mechanical' theorem-proving in GEOMETRY
S/A { [Bk.(Reidel,1988;pp368;CHOU,S.-C.)]

2317. 'Small perturbations' into SUMS OF TWO PRIMES
S/A [MR 19,p393 (VINOGRADOV)]

2318. MODULES as a basis for DIFFERENTIAL GEOMETRY
S/@ [Colloq.Math.24(1971/72)45-79;(SIKORSKI);MR 58 #2845]

2319. APPROXIMATE differentiability (ALMOST-EVERYWHERE)
S/A/@ [PAMS 66(1977)294-298;MR 56 #1291(BRUCKNER/AMS Mem.*)]

2320. PROXIMITY MEASURES for geometric figures
S/A/A [J Cybernetics 2(1972)43-59 (LEE,E.T.);SERRA,J.,Image Analysis(AP,1982)*]

2321. 'Higher matrix COMMUTATIVITY'
δ/@ [Lin.Alg.Applns.2(1969)349-353 (TAUSSKY,O.)]

2322. A 'Gerschgorin theorem' for LINEAR DIFFERENCE EQUATIONS
δ/A/@ [Lin.Alg.Applns.11,27-40;MR:51 #10366;56 #3847(Part II)]

2323. REV.of:'Assigning probabilities to logical formulae'
δ/@ [Synthese 17(4)1967,456-459 (Paper:SCOTT,D./KRAUSS,
    in: Aspects of Inductive Logic(HINTIKKA/SUPPES,eds.)]

2324. ESTIMATES of solutions of sets of LINEAR ALGEBRAIC EQS
δ/A [MR 34 #6583 (CEBAN,V.G.)]

2325. NONLINEAR FUNCTIONAL ANALYSIS:Theory/Applications
δ/A/A [ZEIDLER,E.;5 vols.(Springer,1988-)*--]

2326. Operator th./Anal.Fns./Matrices/ELECTRICAL ENGINEERING
δ/@/A [HELTON,J.W.;Conf.Bd.M Sci.(M) #168(AMS,1987;pp134)]

2327. Spectra of NEARLY-HERMITIAN MATRICES
δ/A [PAMS 48,11-17 (KAHAN,W.);MR 51 #5627]

2328. BLOCK-GERSCHGORIN theorems
δ/@/A [Lin.Alg.Applns.13,45-52;MR 53 #480]

2329. METRICS for INSTANTONS/MONOPOLES
A/A [Phys.Lett.107(1985)21;BAMS(1988)179-183(ATIYAH/HITCHIN)]

2330. Hypotheses EQUIVALENT TO 'RH'
δ/A/A

[PAMS 61245-251(NEWMAN,C.M.);Duke M J (1950)197-]

2331. DISTANCE in ... OPERATOR ALGEBRAS
δ/A/A [Bull.LMS 20,345-349(SOLEL,B.);TAMS(1985)799-817(SOLEL)
MPCPS(1980)327-329(POWER);PLMS(1981)334-356(LANCE)]

2332. Formal DEFINITIONS of ANALOGY/COMPLEXITY
δ/@/A [MR:88f,00022;80i #0370;7,pp123,409;8,p588;10,p132]

2333. Generalizations of the LAPLACE TRANSFORM for PDEs
δ/@ [Ann.Acad.Sci.Fenn.Ser.AI No. 377(1966)(CHURCH,A.!pp34)]

2334. Polynomial BOUNDS for ORDERS (SOLVABILITY)
δ/A/@ [J Algebra (?1977)127-137;234-246]

2335. Functions of n MATRICES:systems of ODEs
δ/@ [ODEs with (quasi-)periodic coeffs.ERUGIN(AP,1966)]

2336. STABILITY problems in NUMBER THEORY
δ/A/A [AMM 73,265-268 (DIXON,J.D.);ULAM,S.,Selected Papers*]

2337. A Complex 'Rolle's Theorem'
δ/@ [AMM 74,452-453 (NEWMAN,D.J.)]

2338. A DISCRETE form of GREEN'S THEOREM
δ/A [IEEE Trans.Pattern Anal. ... PAM 4,242-249(TANG)]

2339. TOPOLOGIES on the space of ALL TORSION THEORIES
δ/A/@ [GOLAN,J:Localization ... (Dekker)*;Pacif.J M 1975+]

2340. Proof of de L'opital's Rule WITHOUT the MVT
       [AMM 76,1051-1053 (BOAS,R.P.)   ]

2341. Comparison of SEPARATION and PROXIMITY spaces
       [AMM 71,158-161 (PERVIN)   ]

2342. Qualitative analysis of NONLINEAR OSCILLATIONS
       [AMS Memoir #244(1981;pp148) (LEVI,M.)*]

2343. BOUNDS on the 'Sup Metric' (probability distributions)]
       [Math.Nachr.Bd.99(1980)95-98 (DALEY,D.J.)   ]

2344. Nearly-uncoupled MARKOV CHAINS
       [Math.Computer Performance(bk)*,287-302(STEWART,G.W.)]

2345. WEBER TRANSFORMS
       [CSIM Lecs.#220*-,143-158 (OLEZIAC)   ]

2346. (i)Universality in Chaos(Papers)
       [CVITANOVIC,P.,ed.(Hilger ,1987+*;now 2nd ed.)]
      (ii)Hamiltonian Dynamical Systems(Papers)
       [McKAY,R./MEISS,J.D.,eds.(Hilger)*   ]

2347. WEAK CONVERGENCE OF PROBABILITY DISTRIBUTIONS
       [BOROVKOV,Stoch. Problems in Queuing Theory(Springer)*]

2348. A function-space METRIC for BV Problems in ELASTICITY
       [Demonstr.Math.9(1976)1-13 (#1892 in ABC[FM])   ]

2349. 'Nine Introductions in Complex Analysis'
δ/A   [SEGAL (N-H,1981;pp716)*--]

2350. REDUCE: Software for Algebraic Computation
δ*   [RAYNA,G.(Springer,1987;pp329)]

2351. APPROXIMATION of the FS for 1/f from the FS for f
A/@   [PAMS 13(1962)965-970]

2352. CV PROOFS of TAUBERIAN theorems(Unified scheme)
δ/@   [MATSCIENCE Report No.56(Madras,1967;pp78 (GAIER,D.)]

2353. TSs 'H-except for X'
A/@   [MR:50 #14662;36 #2107;35 #7299;49 #3800]

2354. Existence of COMPLEXITY GAPS
δ/A/A [JACM 19,158-174(BORODIN,A.);MR 47 #9888]*

2355. ASYMPTOTIC REPRESENTATION of ZEROS of SUM[n^{-z} :n=1...N]
δ/A   [Proc.N Acad.Sci.(USA)985-987(LEVINSON,N.)]

2356. Riemann Zeta Function (recent results)
δ/A   [Bk;Ivic,1983+ ...]

2357. CONVERSION of m-salesmen problems to 1-salesman problems
δ/@   [JACM 21,500-504 (BELLMORE,M./HONG)]

2358. CONSISTENCY/REDUCTION for systems of PDEs (SAM)
δ/@   [JACM 14,45-62 (BRANS,C.H.)]

2359. Theory of computational complexity(overview)
δ/A/A [JACM 18,444-475 (HARTMANIS/HOPCROFT)]

2360. REMOVABLE (null)SETS for analytic functions
δ/A [AMM 75,462-470 (ZALCMAN,L.)]

2361. NEGLIGIBLE FUNCTIONS and Neutrices
δ/A [Indag.Mat.22(1960)115-123* (van der CORPUT)]

2362. SPREAD of a matrix(:=max mod{EV-difference})
δ/A/@ [Duke M J 26(1959)653-661(BRAUER...);HORN/JOHNSON]*

2363. BOUNDED morphisms (in Topos Theory)
δ/A/@ [Topos Theory (CUP,1977)*,JOHNSTON,P.T.;Sec.4.4]

2364. ⎰Geometry of Nonlinear DEs
δ/@ ⎱[HERMANN,R.;Vols.A,B (Math. Sci. Press,1976)]

2365. Extension of Borodin's GAP THEOREM
δ/A/@ [JACM: 19,175-183 (CONSTABLE,R.L.);19,158-174]*

2366. MATRIX CALCULI for computer implementation
δ/A [CACM 13(4)223-237(BAYER,R./WITZGALL)]

2367. COMPLEXITY 'Compression'/'Gap' theorems(Recursion Theory)
δ/A/@ [J Symbolic Logic (1977+)   (JACOBS,B.E.)]

2368. Mechanization of second-order TYPE THEORY
δ/A [JACM 20,333-365* (PIETRZYKOWSKI,T.)]

2369. TOPOLOGICAL criteria for numerical UNSOLVABILITY
δ/A   [JACM 20,399-408 (MILLER,W.)]

2370. Extension of McCarthy's RECURSIVE CALCULUS (LISP)
δ/@   [JACM 20,160-187 (ROSEN,B.K.)]

2371. Structure-preserving morphisms
δ/@   [JACM 19,742-764 (ZIEGLER,B.P.)]

2372. Parallel versions of the FFT algorithm
δ/A/@ [JACM 15,252-264 (PEASE,M.C.);also:Bk(PEASE,AP)]

2373. Prevention of 'Viscious-Circle Definitions'
δ   [Studia Logica t.28(1971)*19-38 (TICHY,P.)]

2374. ANALOGY and DYNAMICAL SYSTEMS
δ/A   [Progr.Cybernetic Syst.Res.8(1982)83-88*(SIEROCKI,I.)]

2375. Algebraic UNIFORMIZATION/parametrization VIA elliptic fns
δ/@/A [Austral.J Phys.27(1974)433-456 (KUMAR,K.)]

2376. ERROR ESTIMATES for matrix equations Ax = b
δ/A   [Duke M J 22(1955)253-261*;Th.43 (BRAUER,A.,et al.)]

2377. Explicit CONSTRUCTION of functions of MATRICES
δ/@   [Rend.Circ.Mat.Palermo Ser.2 t.6(1957)*(LEHRER,Y.)]

2378. Homotopy theory of MODULES
δ/@

[MR 20 #4588 (HILTON,P.J.)]

2379. Tauberian REMAINDER THEOREMS
S/A [LNM #232 (GANELIUS,T.);recent papers*-]

2380. Uncommon results in MATRIX THEORY
S/@ [Els.of Matrix Theory (MEHTA,M.L;HindustanPubl.Co.,1977)]

2381. 'Incorrect'algorithms still useful for 'meaningful inputs'
S/@ [LNCS #53*,p148- (cf asymptotic series ...)]

2382. DISTANCES on general SIGNAL SETS
S/A/A [Progr.Cyber.8(ed.TRAPPL,et al,.pp213216;GROLLMANN)]*

2383. (I)KBS for TSs
S/A [Progr.Cyber.8(TRAPPL,ed.)263-276(PASSOS,E.P.L.)]

2384. DISTANCE of a PRIME IDEAL from a MODULE
A/@ [Bull.LMS (3/79)33-36 (DUTTON,P.)]

2385. FORMULATION of differential geometry VIA MODULES
S/A [?MR 58 ... ?(LMS LN (KOCK,A.)*)]

2386. Linear algebraic aspects of NETWORK THEORY
S/A [Classical N-W Th (BELEVITCH,V.(H-D,1968);MR 39 #5243]

2387. FIXED-POINT Theory for n-connected manifolds
S/A/@ [Topology 20(1978)53-92(FADELL,E./HUSSEINI,S.]*

2388. Ordinary/PARTIAL FUNCTIONAL EQUATIONS
δ/@ [MR 30 #5073(KUCZMA,M.);?also bk(KUCZMA)]

2389. ⎧ OPERATOR INEQUALITIES (Matrices,O/PDOprs ...)
δ/A/@ ⎩ [SCHRODER,J.,bk.(AP,1980;pp367)]

2390. PATH-INSENSITIVITY semigroups
δ/@/A [Arch.Mech.Stos.26(1)(1974)119-133 {ABC(MFM)$\widetilde{\#}$1545*}]

2391. TORSION of spaces(Cartan)/DISLOCATIONS
δ/A [Mech.Generalized Continua(KRONER,ed.)*]

2392. MEASURE OF NONMONOTONICITY of (arithmetic) functions
δ/A/@ [Pacif.J M 77(1978)83-101 (DIAMOND,H.G./ERDOS,P.)]

2393. ANALOGUES(for {max.cgt.polynomials}) of JENTZSCH's Theorem
δ/A/A [Duke M J 26(1959)*605-616]

2394. 'Hilbert METRIC' in BSs
A/@ [e.g,FP-Th.(bk); ISTRATESCU,V.I.(Reidel,1981;Sec.3.4*)]

2395. NONSTANDARD ANALYSIS/'MULTIPLICATION' of GFs
δ/@ [Scientia Sinica 21(1978)561-585*]

2396. UNIVALENCE properties of{ qf+(1-q)g }
δ [Rocky Mountain J M 15(1975)475-492(CAMPBELL,D.M.)]

2397. ANTICONVEX SETS
δ/@

[Rev.Roum.M P&A 13(1968)1399-1401 (MARCUS,S.)]

2398. { Sommes d'Exponentielles }
δ/A [SCHWARTZ,L.'bk*;MR 21 #5116(KOOSIS,P.),long REV.]

2399. { Symbolic Operators }(Direct justifcn of Heaviside calc.)
δ [DALTON,J.P.(bk);Witwatersrand U P,1954;M Gaz.1955*,255-]

2400. CONVERSE(s) to the CONTRACTION-MAPPING THEOREM
δ/@ [Usp.Mat.Nauk 31(1976)169-198/Russ.M Surveys?*]

2401. INfinite-element method(Numer.soln.opr.eqs.)
δ/A/@ [Numer.Math.39,39-50(HOUDE HAN);papers*-]

2402. { Dictionary of Physics/Maths.ABBREVIATIONS }
δ [POLON,D.(ed.);ODDYSSEY(NY,1965);pp333   (?Inst.)]

2403. ITERATIVE SOLUTION of Wiener-Hopf equations
δ/A [QAM 20,341-352 (WUs,T.T.)]

2404. ALMOST-POSITIVE matrices
δ/@ [DONOGHUE,Grundl.#207(Springer,1974) (?Inst.)]

2405. Generalizations of Gerschgorin's theorem
δ/A/@ [Pacif.J M 12(1962)1241-1250 (FEINGOLD/VARGA)]

2406. BOUNDS on functions of matrices
A/@ [AMM 74,920-926*(ROSENBLOOM,P.C.)]

2407. CATEGORICAL APPROXIMATION(Shape Theory)
δ/A/@ [Bk;CORDIER,J-M,et al.(eds.) (Wiley/Horwood,1989;pp207)]

2408. INfinite-dimensional versions of SARD's Theorem
δ/A/@ [Amer.J M 87(1965)861-866?* (SMALE,S.)]

2409. GENERALIZED (D)AC
δ/@ [Mat.Sbornik 76(118)(1968)135-146;MR 38 #323]

2410. Properties of p- and (p,q)- analytic functions
δ/@ [POLOZII;bk.;MR 34 #357]

2411. NULL-SETS for classes of analytic functions
δ/A [AMM 75,462-470*(ZALCMAN,L.)]

2412. ASYMPTOTIC FORMS of 2-D FTs
δ/A/@ [Duke M J 27(1960)581-596 (DUFFIN,R.J./SHAFFER,D.H.)*]

2413. ACCUMULABILITY(a generalization of SUMMABILITY)
δ/A/@ [Duke M J 27,555-560 (POSNER,E.C.)]

2414. FINITE LTs
δ/@ [MPCPS 63,155-160 (DUNN,H.S.);several Bks*-]

2415. Special Fuctions of MATRICES
δ/@ [The H-Function (MATHAI/SAXENA;Wiley,1978)*]

2416. Global optimization VIA SIMULATED ANNEALING
δ/A/A

[Bk*;papers*--]

2417. Topologization of sequence spaces/Toeplitz matrices
δ/A/@ [Michigan M J 5(1958)139-148 (ERDOS,P.,et al.)]

2418. Sequence spaces (general treatment;BSs,etc.)
δ/@ [RUCKLE (Pitman Res.N M) (?Inst.)]

2419. Abstract Hilbert trans./related semigroups/PDEs
δ/@ [Res. N M #8 (paper by DETTMAN)*]

2420. Integration-by-parts for CESARO-SUMMABLE INTEGRALS
δ/@ [JLMS 29(1954)276-292 (BORWEIN,D.)*]

2421. EXTNs.(to R(n)/other PDOs)of Cauchy's Integral Formula
δ/@ [JLMS 21(1946)210-218* (WEISS,P.)]

2422. 'Paley-Wiener Theorems' for RADON TRANSFORMS
δ/@ [Comm.P&A M 23,409-424* (LAX,M./PHILLIPS,R.)]

2423. Spaces(of functions) defined VIA LOCAL APPROXIMATIONS
δ/A/@ [Trans.Moscow M S 24(1971)*73-139 (BRUDNYI)]

2424. LOCAL theorems in UNIVERSAL ALGEBRAS
δ/A/@ [JLMS 34(1959)177-184(McLAIN)]

2425. Generalization of MEANS in classical analysis
δ/A/@ [J Combin Inf Syst Sci 3(1978)175-199*;MR 80d 26017;
        (BRENNER,J.L)]

2426. ANALYTICITY in Linear Algebras
S/@ [Duke M J 27,431-441* (TRAMPUS,A.)]

2427. STABILITY properties of GEOMETRIC INEQUALITIES
S/A [AMM 97,382-394(GROEMER,H.)*]

2428. Abstract treatments of HYSTERESIS
S/@/A [KRASNOSEL'SKII/POKROVSKII,(bk)* (Springer,1988)]

2429. SIMULTANEOUS (D)AC of general 'objects'
S/@ [JLMS 34,264-272 (EDWARDS,R.E.) }

2430. Function-space analogues of sequence-space results
S/A [QJM(Ox)11(1960)310-320 (PRASAD,S.N.)]

2431. BOUNDARIES induced by nonnegative MATRICES
S/A/A [TAMS 83(1956)19-54;Ann.Math.65(1957)527-570(FELLER)]

2432. ACCELERATION of stochastic APPROXIMATION
A/A [Ann.Math.Stat.29(1958)41-59 (KESTEN)]

2433. DISTINGUISHABILITY of distributions(Prob./Stats.)
S/A [Ann Math Stats 29,700-718 (HOEFFDING/WOLFOWITZ)]

2434. Asymptotic convergence to the NORMAL DISTRIBUTION
S/A [Ann Math Stats 29,373-405 (SACKS,J.)]

2435. INTERPOLATORY Fn.Th.(reduction of double integrals)
δ/A [J Approx.Th.5(1972)276-307 (DAVIS,P.J.)]

2436. Nonlinear Volterra integral eqs. AS CONTRACTIONS /bk*
δ/A [ARMA 15(1964)79-86 (WILLET,D.)]   (Whittaker)

2437. Operators WITHIN REACH of ... an ALGEBRA
δ/A/A [LMS LN #76* (CORDES,H.O.)]

2438. ESSENTIAL DEPENDENCE for functions of n variables
δ/@ [JLMS 41,333-335 (DAVIES,R.O.)]

2439. N-diml.EXTNS. of Toeplitz-determinant INEQUALITIES
δ/A/@ [Math.USSR Izv.9(195)(6)1323-1332 (LINNIK)*]

2440. Approximation of functions in the HAUSDORFF METRIC
δ/A [Math USSR Sb.30(1976)449-477 (JU,I.)]

2441. SOLVABILITY of arbitrary equations in free SEMIGROUPS
δ/@ [Math.USSR Sb.32(1977)129-198 (MAKANIN)]*

2442. Singular integral eqs./matched expansions,for DISLOCATIONS
δ/A/A [Math USSR Sb.34,475-502 (NOVOKSENOV)]

2443. Generalizations/extensions of CHOQUET THEORY
δ/@ [Russ.M Surveys 30(1975)115-155 (KUTATELADZE)]

2444. Proximity INVARIANTS
δ/A [Soviet M Dokl.15(1974)24-28 (EFREMOVIC,et al.)]

2445. Generalized CONTRACTION MAPS
δ/A/@ [Soviet M Dokl. 15,673-676 (GORBUNOV)]

2446. ANALOGUES of the KOROVKIN Approximation Theorems
A/A [Soviet M Dokl.15,1433-1436 (GADZIEV,A.D.)]

2447. Morse Theory and DOMAINS OF ATTRACTION
δ/A [Rend.Circ.Mat.Palermo13(1964)229-238 (LEIGHTON,W.)]

2448. Solution of generalized infinite sets of linear alg.eqs.
δ/A/@ [Rend.Circ.Mat.Palermo(RCMP) 11,5-24 (MARCUS,B.)]

2449. Analysis of diffeo.approxns.to C(m)-diffeomorphisms
δ/A/@ [RCMP:11,25-46;291-318 (HUEBSCH/MORSE)]

2450. q-Integrals /q-Laplace Transforms
δ/@ [RCMP 11,245-261;Math.Z.(1962+):ABDI]

2451. Differentiation(/(D)AC) on(noncommutative) ALGEBRAS
δ/@ [RCMP 11,204-216 (RINEHART,R.F./WILSON,J.C.)]

2452. NONLINEAR extns.of the Heaviside OPERATIONAL CALCULUS
δ/@ [LAMNABI,et al.;Preprint*/Antwerp Conf.(SAM)]

2453. Nonlinear vector integral eqs.AS CONTRACTIONS
δ/A [ARMA(1964) (WILLET,D.)*]

2454. Algebraic-geom. formulation of n-SINGULAR PERTURBATIONS
δ/A/A [TAMS 156(1971)1-43 (STENGLE,G.)*]

2455. QUASI-INVERSE operators
δ/@ [MASLOV,V.,P.:Operational Methods (MIR,1977)*]

2456. Prime decomposition of AUTOMATA
δ/A [Math.Syst.Th.11(1978)239-257* (DORFLER,W.)]

2457. (Musical) PITCH-STRUCTURES AS order-preserving MAPS
δ/A [Math.Syst.Th.11,199-234 (ROTHENBERG,D.)]

2458. Discrete forms of 'GREEN'S THEOREM'
δ/@ [IEEE Trans.Pattern Anal.(5/82) (TANG,G.Y.)]

2459. RESPONSE MAPS over rings
δ/A [Math.Syst.Th.?11(?1977)169-175 (SONTAG,E.D.)]

2460. BOUNDS for classes of DETERMINANTS
δ/A [Duke M J 22(1955)95-102 (OSTROWSKI,A.M.)]

2461. QUASI-STOCHASTIC MATRICES
δ/@ [Duke M J 22,15-24 (HAYNSWORTH,E.V.)]

2462. RANDOM 'Tietze',and,'Heine-Borel' Theorems
δ/@/A [Ann.Math.Stat.30(1959)1152-1157 (HANS,O.)]

2463. GLOBAL ANALOGUES:Diophantine Approxn./Value-Distrbn.Th.
δ/A/A [LNM #1239* (VOJTA,P.A.)]

2464. Structure of K if K+K is 'small'
  S/A/@ [LNM #1240 (FREIMAN,G.)]

2465. Uniformization VIA Elliptic Functions ...
  S/A [Austral.J.Phys.27(1974)433-456(Appendix);(KUMAR,K.)]

2466. Topological CONTIGUITY
  S/A/@ [SPANIER,E.:Algebraic Topology,(bk)*,p 130-]

2467. Proof that:Structurally stable systems are NOT DENSE
  S/A/@ [Amer.J M 88(1966)491-496 (SMALE,S.)]

2468. 'Convexity-generalizations' of basic INEQUALITIES
  A/@ [QJM(Ox)40,247-250;PCPS 64,1023-1027(DAYKIN/ELIEZER)]

2469. NULL-SETS of analytic functions
  S/A [AMM 75,462-470* (ZALCMAN,L.)]

2470. 'KOROVKIN Theorems' related to the HAUSDORFF METRIC
  S/@ [Soviet M Dokl.(1967)1445-?1557 (SENDOV)]

2471. Diverse EXAMPLES of UNIFORM ALGEBRAS
  S [Th. of Uniform Alg. (STOUT,E.L.);Bogden&Quigley,1971]

2472. Differentiability of functions over ALGEBRAS
  S/@ [RCMP 11(1962)204-216 (RINEHART/WILSON)]

2473. PRESERVATION of properties of LCTVs
  S/@ [L N P&A M (Dekker,1976) (McKENNON,K.,et al.)]

2474. (Partial)CONVERSES of the Hadamard Product Theorem
S/A/@ [Duke M J 26,133-136 (BUCK,R.C.)]

2475. The LAMBERT TRANSFORM and its INVERSE
S/A [Duke M J 27,561-568 (PENNINGTON,W.)]

2476. Transl.-approxn.UNIVERSALITY of Zeta-fns/L-fns
S/@ [M USSR Izv.9(1975)443-453 (VORONIN)]

2477. HAUSDORFF-APPROXIMATION by piecewise-MONOTONIC functions
S/A [M USSR Sb.30(1976)449-477 (DOLZENKO,et al.)]

2478. Singular integral equations in DISLOCATION THEORY
S/A [M Sb 34(1978)475-502 (NOVOKSENOV)]

2479. { Les Methodes Nouvelles de Mec.Celeste }
S [English TRANSL.(Hilger,1990+;4 Vols.(B's)]

2480. INFINITE-ELEMENT METHOD for interfaces
A/@ [Numer.Math.39,39-50 (HAN,H.)]

2481. CONSTRUCTIVE solution of POLYNOMIAL EQUATIONS
S/A [Soviet M Dokl.20(1979)166-169 (MINASJAN)]

2482. APPROXIMATE IDENTITIES in NORMED ALGEBRAS
S/A/@ [JLMS 17,141-151;PLMS 26,485-496:(DIXON,P.G.)]

2483. EVOLUTIONARY DISTANCE
A/A [J.Algorithms 1(1980)359-373 (SELLERS,P.H.)]

2484. IRREDUCIBILITY in CS control structures
S/A [CACM 18(1975)629-639 (LEDGARD/MARCOTTY)]

2485. ALMOST-FIXED-POINT Theory
A/@ [Canad. M J 30(1978)(HAZEWINKEL,et al.;vd VEL;M Centrum]

2486. Computer-aided proofs in Analysis
S [MEYER,K.R./SCHMIDT,eds.(IMA/Springer,1990;pp265)]

2487. COMPUTABILITY in Analysis and Physics
S/A/A [Perspec.M Logic (POUR-EL/RICHARDS,eds.);Springer,c1990)]

2488. A 'natural METRIC' for CAFs
S/A [AMM 65(1958)756-758 (GLEASON,A.M.);MR 20 #7092]

2489. Polar (asymptotic)decomposition of JORDAN MATRICES
S/A [J R A(ngew.)M 200(1958)190-199(OSTROWSKI);MR 20 #7036]

2490. METRIC SPACES of thermodynamic states
S/A/A [e.g.,LANDSBERG;BUCHDAHL (bks)*]

2491. Algebraic proofs of Kirchhoff's NETWORK THEOREMS
S/A [AMM 68(1961)244-247 (NERODE/SHANK)]

2492. Continuous ANALOGUES of SERIES
S/A/A [AMM 80,18- (BOAS,et al.)]

2493. Generalized GREEN'S FUNCTIONS/matrices
δ/@   [SIAM Rev.12(1970) (LOUD,W.S.)]

2494. Max.-mod.(asymp.)E-V formulation of St Venant Principle
δ/A/A [CHEREPANOV,bk(MFM)*- p45-]

2495. Extensions of 'Cauchy integral formula'VIA stress tensors
δ/A   [JLMS 21,210-219 (WEISS,P.)]

2496. Generalizations of the KRONECKER PRODUCT
δ/@   [Bull.Calcutta M S 73(1981)295-306 (MAULIK,et al.)]

2497. SUPERCONVEXITY of 'max.-mod. eigenvalues'
δ/A   [QJM(Ox)12(1961)283-284 (KINGMAN,J.F.C.)]

2498. Normal modes in NONlinear systems
δ/@   [J Appl.Mech.29(1962)7-14 (ROSENBERG,R.M.);MR 25 #792]

2499. APPROXIMATION theorems for TRANSLATES
δ/A   [PLMS 9(1959)321-342 (EDWARDS,R.E.)]

2500. TOPOLOGIZATION of arbitrary sets VIA abstract distance-fns
δ/A/@ [MAMUZIC,Z. GT(Noordhoff,1963)*,Sec.12]

2501. General STRUCTURES in mathematics
δ/@   [BOURBAKI,Set Theory (AP)*--]

2502. Measures/metrics in 2-valued LOGICS
A/@   [Mathematika:Rev.d'Anal.Numer.et ... 20(43)113-118(BOTH)]

2503. LACUNAE in 2-D WAVE PROPAGATION
δ/A   [PCPS 63,819-825 (BURRIDGE,R.)]

2504. Additive(but NOT contably-additive)SET-Functions
δ/@   [PCPS 63767-775 (KINGMAN,J.F.C.)]

2505. SELF-RECIPROCAL SERIES
δ   [PCPS 68,427-438 (BILLINGTON);TITCHMARSH (bk)*]

2506. Operators CONSERVING positive-definiteness
δ/@   [PCPS 69,87-97 (RIDER,D.)]

2507. Conservation of #(eqs) under algbr.-geom.SPECIALIZATION
δ/@   [PCPS 69,59-70 (NORTHCOTT,D.G.)]

2508. NONLINEAR transformations of infinite SERIES
δ/@   [J Res.NBS 73B(1969)251-274* (GRAY,H.L./CLARK,W.D.)]

2509. Quasi-ISOMETRIC measures
δ/A/@ [MR 41 #3704]

2510. For GROUPS: G-solvable --> G-measurable
δ   [Theorem of v NEUMANN (?REF)]

2511. LARGE subgroups of semi-simple LIE GROUPS
δ/A   [Grundl.bd 188,p305- (WARNER,G.)]

2512. Positive-def.n-forms NOT represbl.as wtd.sums of squares
δ   [Proc.Conf.Q U.(Ontario)(1976)385-405;MR 58 #16503]

2513. GENERIC properties in Topological Dynamics
   [LNM #468,241-250;MR 58 #31268 (PALIS,J.,et al.)]

2514. 2-METRICS and 2-NORMED SPACES
   [M Nachr.91(1979)*151-155 (RHOADES,B.)]

2515. Quasi-ANALYTIC functions
   [Bull.Calcutta M S 39(1947)157-165 (ZAHORSKI,Z.)]

2516. Converses of the Banach FP Theorem
   [PAMS 18(1967)287-289 (LUDVIK,J.) MR 34 #8398]

2517. Continuum theories of DISLOCATIONS
   [KRONER,E.(ed.),bk;MR 20 #2117]

2518. ALMOST-hypoelliptic differential operators
   [PCPS 67,283-293;PLMS 39(1969)537-552 (ELLIOTT,R.J.)]

2519. The LAMBERT TRANSFORM (and its INVERSE)
   [Duke M J 27,561-568 (PENNINGTON,W.P.)]

2520. Iterative solution of Wiener-Hopf equations
   [QAM 20,341-352 (WUs,T.T.)]

2521. Electrical ANALOGUES of stochastic processes
   [MPCPS 80,145-151 (KELLY,F.P.)]

2522. Stochastic EXTENSIONS of (ill-posed) linear problems
δ/A/A   [J M Anal.Appl.31(1970)682-716 (FRANKLIN,J.N.)]

2523. Two-variable Laplace TRANSFORMS
δ/@   [VOELKER/DOETSCH,bk(Birkhauser,1950)] / Dtkn-Prdnkv*

2524. Matrix LOGIC(positive/negative/hypercube/continuous/...)
δ/A   [STERN,A.,bk(N-H,1988;pp250)]

2525. INfinite Legendre integral TRANSFORMS
δ/@   [PCPS 57,546-560 (CLEMMOW,P.C.)]

2526. Integral IEQUALITIES involving f and f'
δ/A   [AMM 78,705-741 (BEESACK,P.R.)]

2527. Material CLOSENESS (VIA conductivity tensors)
δ/A/A   [KNOPS/LACEY,eds.,LMS LN #122*p141-   (TARTAR)]

2528. INSTABILITY of the CONVERSE of T for 'small changes in T'
δ/A/A   [ROSSER,J.B.,Logic for Mathematicians(McG-H,1953)*,p45-]

2529. Generalized (D)AC/Unique meromorphic extensions
δ/@   [Mat.Sb.76(1968)135-146* ;MR 38 #323]

2530. { Perspectives in NONLINEAR DYNAMICS }
δ/@   [(CUP,1990+;2 vols.); JACKSON,E.A.]

2531. Baxter's 8-Vertex Solution withOUT elliptic functions
δ/A   [Austral.J.Phys. 27(1974)433-456 (KUMAR,K.)]

2532. Nonlinear oscillations/ANALOGUES of Riemann surfaces
S/A/A   [Proc.4th Conf.Nonlin.Oscl.(Prague,1967),p167(Abstr.)]

2533. EXTENSION of the Kontorovich-Lebedev transform to GFs
S/@   [MPCPS 77,139-143 (ZEMANIAN)]

2534. Functional-analytic proof of Rouche's Theorem
S/A   [AMM 78,770-   (v DULST,D.)]

SiMA/MaAT:MAIN LIST of TOPICS

1. CONTINUITY

2. DIFFERENTIABILITY

3. ANALYTICITY

4. BOUNDEDNESS (/BOUNDARIES)

5. UNIFORMITY

6. COMMUTATIVITY

7. EMBEDDABILITY

8. LINEARITY

9. INEQUALITY

10. CONSTRUCTIBILITY

11. DENSENESS

12. DENSITY

13. TRANSCENDENTALITY

14. INVARIANCE
15. APPROXIMABILITY
16. PERIODICITY
17. INTEGRABILITY
18. STABILITY
19. DUALITY
20. CONNECTEDNESS
21. ACCESSIBILITY
22. APPLICABILITY
23. DEPENDENCE
24. EXTREMALITY

25. INDETERMINACY
26. REPRODUCIBILITY
27. REPRESENTABILITY
28. REPLAC(E)ABILITY
29. COMPLEXITY
30. SINGULARITY
31. UNIVERSALITY
32. ADDITIVITY
33. CONVERGENCE
34. COMPACTNESS
35. INVERTIBILITY
36. ITERABILITY/ITERATION
37. METRIZABILITY/NORMABILITY/NON-NUMERICAL 'DISTANCE'
38. DISTINGUISHABILITY
39. DECOMPOSABILITY
40. PROPAGATABILITY/PROPAGATION
41. DEFORMABILITY

42. RECONSTRUCTABILITY/RETRIEVABILITY

43. EFFICIENCY

44. SEPARABILITY/INTERACTION

45. ORDERABILITY/ORIENTABILITY

46. INTERFACE ANALYSIS

47. INTERFERENCE

48. ASSOCIABILITY/ASSOCIATION

49. REDUCIBILITY

50. CONSTRAINABILITY/CONSTRAINT

51. FINITENESS

52. CONVEXITY

53. INFECTABILITY/CONTAGION

54. EXTENDABILITY/EXTENSION

55. COMPLETENESS

56. POSITIVITY/POSITIVE-DEFINITENESS

57. CHARACTERIZABILITY/CHARACTERIZATION

58. COMPATIBILITY/COMPARABILITY

59. ERGODICITY

60. BIFURCATION

61. LOCALIZABILITY

62. RESPONSE

63. SYMMETRIZABILITY/SYMMETRY

2535. δ/A **Potential of anisotropy** (VOLKOV, SD, Sov. Phys. Dokl. 16 (1971) 547–9; MR 47 #2886) [Reduction of **an**iso. elast. probl. to **iso** probl. + 'Poisson' PDE.]

2536. $\delta/@$ **Discrete AC for solutions of differential equations** (JM Sn. Appl. 9 (1964) 252–267, DUFFIN / DURIS MR 30 #374)

2537. $\delta/@$ **Extension of Cauchy $\int$-Formula to ∞-conn. dom.** (MR 47 #2882)

2538. $\delta/@$ **'Green's Functions' for difference equations** (MR 21 #7373 Proc. M/Ph Soc Egypt 22 (1958) 43–51, MISSIH)

2539. δ/A **Prob $[\{t_n\}$ is a basis for $\mathbb{P}]$** (Math. Scand. 4 (1956) 303–8; L Carleson)

2540. $\delta/\mathcal{A}$ **Accuracy of 'regularization' for $\int$-eq.** (Math. Comp. 28 (1974) 889–907; FRANKLIN, JN)

2541. $\delta/@$ **Quasi-coordinates** (Whittaker: Anal. Dyn. (bk)* White$^+$ Elec/Mech. Energy Convsn. (bk)* – modified Lagrangians)

2542. $\delta/\mathcal{A}/@$ **Topologization of Countable Algebras** (MR 80d 08003 TAIMANOV (!))

2543. $\mathcal{A}/A$ **Dense approximants for elast. wedges** (J. Appl. Mech. 32 (1965) 26–30/MR 30 #3906)

2544. δ/A **Turnpikes 'as' attractors (?)** (e.g. TAKAYAMA, Math Econ. (bk)*)

2545. δ/A **Ideals 'as' functions of 'base rings'** (Sympos. Math XI* (1973) 105–120 NORTHCOTT)

2546. $\delta/\mathcal{A}/@$ **Distance in operator algebras** (BLMS 20 (1988) 345–9 Baruch SOLEL)

2547. δ **'Real Analysis and Probability' by RM DUDLEY** (C&H pp.436, (1989) → BLMS 22 (4) (1990) 411–2)

2548. $\delta/@$ **Cardinality of Structures in Complete 1st-order classes** (BLMS 19 (1987) 209–237, Wilf H.) ('All or nothing' results ...)

2549. $\delta/@$ **Galois Theory for partial algebras** (LNM #1004, 257–272, ROSENBERG)

2550. $\delta/\mathcal{A}/A$ **Spectral sequences 'as' approximations to 'limit modules'** (e.g. SPANIER, Alg. Topol.* 466ff)

2551. $\delta/\mathcal{A}/A$ **Difference-analogues of singular $\int$-eqs.** (Trans. Moscow MS (1984) (2) 43–67, BELYANKOV, et al.)

2552. δ/A **'Blocking Coalitions' (games) vs. 'Unavoidable Configurations'** (refs. rqd!)

2553. $\mathcal{A}/@$ **(Strongly) negligible sets** (BLMS 19 (1987) 371–7, DIJKSTRA, J)

2554. $\delta/\mathcal{A}/A$ **Mahalanobis distance (stats.)** (e.g. Topics Adv. Econometrics (DHRYMES)* Vol. 2 p31 (Springer, 1994))

2555. $\delta/A/@$ **Materials with 'manifold-based microstructures'** (Mono. CAPRIZ, Springer, 1988) (See SIAM Rev. for brief account)

2556. $\delta/@$ **Almost-similar operators (spectra...)** (TRANS. MOSCOW MS 1979 (2) 57–82; ROZENBLJUM)

2557. $\delta/@$ **Hodge-Th./Alg. Topol. (variations)** (PRSLA (1984) (WALL, CTC))

2558. $\delta/@$ **Sharpened RADON transform** (PRSL A422, 343–9 MOSES/PROSSER)

2559. $\delta/\mathcal{A}/A$ **Pseudo-nonlinearities in engineering structures** (PRSL A445, 193–220 WORDEN, K^+)

2560. δ/A **Complex representation (by 6 'potentials') of 3-D elast. solutions** (J. Elast. 22 (1989) 45–55/18, 191–225; PILTNER) [Also: Acta. Mech. 75 (1988) 77–91; Mech. Res. Comm. 15 (1988) 79–85]

2561. $\delta/\mathcal{A}/A$ **'Geom.' vs. 'Anal.' closeness (Whitney norm)** (Geom. μ-Th. H FEDERER)

2562. $\delta/\mathcal{A}/A$ **Directed graphs / stability / chem. reactions** (PRSL A334, 299–342 I, II, III: F. HORN) [Also: ARMA 49, 172–186 / GLANSDORFF-PRIGOGINE (bk)*]

2563. $\delta/\mathcal{A}/A$ **Perturbations of BVP (stability)** (BULL Acad. Polon. Sci. II (1963) 23–26; K MAURIN)

2564. δ/A **Transmission lines / basic QM (analogues)** (Elec./Holes in Semi-cond. SHOCKLEY (bk)* V. Nostrand, 1950)

2565. $\delta/\mathcal{A}/@$ **Block-Gerschgorin criterion (e.-v.)** (MR 53 #480)

2566. $\delta/@$ **Identity theorems for maps in TS** (MR 37 #2163 (MA-MUZIĆ))

2567. $\delta/\mathcal{A}/A$ **'Continuous Ctd. Fractions'** (See WALL, HS 'Ctd. Fractions' (bk)*)

2568. $\delta/\mathcal{A}/@$ **Systematic approximation of GF** (TREVES, F 'TVS, Distributions & Kernels' (bk)* Ch. 28)

2569. $\delta/\mathcal{A}/A$ **Various 'distances' for point processes** (REISS, RD, 'A course on Pt. Proc.', Springer, 1993*)

2570. $\delta/\mathcal{A}/A$ **Anisotropic Elasticity (Stroh $(+)^n$)** (TCT TING, OUP, 1996)

2571. $\simeq$ 2547. (δ)

2572. $\delta/\mathcal{A}/A$ **Splitting fields for (Gaussian) Stoch. Proc.** (ARMA 43 (1971) 367–391; L PITT) [Also: see 'Repro.-kernel HS' (Weinert, ed. ?*)]

2573. $\mathcal{A}/@$ **'Grothendieck Spaces in Approximation Theory'** (Mem. AMS #120 (pp121) BLATTER, J)

2574. $\mathcal{A}, A$ **Stochastic distances** ('Detection & Estimation' (CSP, 1990), (KAZAKOS)^2, Ch. 11)

2575. δ/A **Stress / Strain / Structural Matrices** (PILKEY, WD (Wiley, 1994) pp. 1458!)

2576. $\delta/\mathcal{A}/@$ 'Contraction outside a compact set $\ni fp$ (AMM 76 (1969) 565 Problem 5672)

2577. $\delta/\mathcal{A}/@$ **Measures in Countable spaces** (AMM 76 (1969) 494–502 HANISCH / HIRSCH)

2578. $\delta/\mathcal{A}/A$ **Quasi-city / order-topol. / (unif.) approximation** (AMM 76, 489–494, JA DYER, AMM 73, 144–5)

2579. $\mathcal{A}/@$ **Approximation of smooth manifolds by real algebraic sets** (Russ. Math. Surveys 37(1), 1982; 1–59)

2580. $\delta/\mathcal{A}/A$ **Hankel operators / best approximation / Gaussian processes** (Russ. Math. Surveys 37(1), 1982, 61–144)

2581. $\delta/\mathcal{A}$ **'Inverse problems for ODE'** (Yu. S / OSIPOV / KRYAZHINSKII bk (G&B (1995), pp625) (Highly constructive))

2582. = 2575. (δ/A)

2583. $\delta/\mathcal{A}/@$ **FP-Theorems for multi-values functions** (AMM 73 (4) (1966) 351–355 RE SMITHSON)

2584. $\delta/@$ **Preservation of proportions under** $\{\tau_\alpha : \alpha \in \Delta\} \to \prod_\alpha \tau_\alpha$ (AMM 73, 358–360, N LEVINE) (Here, the τ_α are topologies on a fixed set X)

2585. δ/A **Statistical isomorphism** (Ann. Math. Stats. 37 (1966) 203–214 N MORSE$^+$) ($\ni$ Cat.-th. formulation...)

2586. $\delta/\mathcal{A}$ **A Compendium of nonlinear ODE** (PL SACHDEV (Wiley, 1997) pp.800! ODE of order ≤ 5)

2587. δ/A **Use of Matroids in Structural Mechanics** (PRSL A350, 61–70, CASSELL, AC +)

2588. $\delta/\mathcal{A}/A$ **Local Surface Morphology / entire functions** (PRSL A360, 25–45 G ROSS +)

2589. $\delta/@$ **Kontorovich-Lebedev transformations / distributions** (PCPS 77 (1975) 139–143 ZERMANIAN)

2590. $\delta/\mathcal{A}/A$ **Double-series solution of biharmonic eq.** (PCPS 76 (1974) 563–585 VAUGHAN, H)

2591. $\mathcal{A}/@$ **Abstract distances in groups** (MPCPS 80 (1976) 451–463 CHISWELL, IM)

2592. δ **Equivalence of COMPACT and FINITE in TS** (AMM 75 (1968) 178–180; LEVINE, N)

2593. $\mathcal{A}/A$ **'Circuit-th.' as an approximation to 'field equations'** ('EM Waves & Radiating Systems' JORDAN / BALMIN, P-H, 1968*, §14.16)

2594. = 2585. (δ/A)

2595. $\mathcal{A}/@$ **Approximation of bi-commutant (M″) of M by M** ('VS' FS CATER* p99 (Saunders))

2596. $\mathcal{A}/A$ **Cts analogues of series** (AMM 80 (1973) 18 BOAS / POLLARD)

2597. δ/A **Applications of SAM to Dynamics** (HC HOWARD, Tech. Press, 1980, pp.404)

2598. = 2547. (δ)

2599. $\mathcal{A}/A$ **Skorohod / Prohorov-Cgce of probability distributions** (AA BOROVKOV*... Queueing Th., Springer, 1976, p.25)

2600. $\delta/\mathcal{A}/A$ **Stability of maps / statistical distributions** (PRSL A415 (1988) 445–452 MK MURRAY)

2601. $\delta/\mathcal{A}/@$ k-**dimensional Ctd. fractions** (MR 47 #1753 G SZEKERES)

2602. $\delta/\mathcal{A}/A$ **Piecewise-lin. th. of nonlinear NW** (SIAM J. Appl. Math. 22 (1972) 307–) (Are 'almost all circuits' stable?)

2603. δ/A **Chem. applications of topol. / graph-th.** (ELSEVIER, 1983*, RB KING, ed.)

2604. δ/A **Crystal Structure / Information Th.** (Proc. National Acad. Sci. USA 44 (1958) 948–956) [See MR 20 #2932 DA McClachlan: diffraction]

2605. δ/A **Axomatization of structural Chem. Models** (MULLKHUYSE, JJ, Thesis A'dam, 1960, pp/67. MR 22 B #7775)

2606. $\delta/@$ **LT-Calculus for PDE** (Mem. AMS #143 (1974) T DONALDSON MR 58 #29483)

2607. δ/A **The Modified Riemann S. 'as' the Fundamental Group** (MR41 #3043 T REGGE)

2608. $\mathcal{A}/@$ **Iterative solution of inf. syst. poly. eqs.** (Rend. Circ. Mat. Palermo (II) XI (1962) 5–24, MARCUS, B)

2609. $\mathcal{A}/@$ **Approximation of C^m-diffeos by ANALYTIC diffeos** (Ibid. 25–46 HUEBSCH / M MORSE)

2610. $\mathcal{A}/A$ **Extremal problems for simplicial complexes** (Ibid. 179–184 LC YOUNG: +II–VI...)

2611. δ/A **Boolean geom. / TS** (Rend. Circ. Mat. Palermo (1952) 343–360; (1961) 175–192 BLUMENTHAL / PENNING)

2612. $\delta/\mathcal{A}/@$ **Syntopogenous pre-ordered spaces** (PCPS 80 (1976) 71–79 D. BURGESS)

2613. $\delta/@$ **Eigenvalues of composite matrices** (PCPS 57 (1961) 37–49, B FRIEDMAN)

2614. δ **Singularity-propagation for PDE** (J. Fac. Sci. U. Tokyo* 37(2) (1990) 377–424 ISHII, T)

2615. δ/A **Stochastic Ctd. Fractions** (Rend. Circ. Mat. Palerno (1952) 170–208 (Paul LÉVY))

2616. = 2567. ($\delta/\mathcal{A}/A$)

2617. $\delta/@$ **Rigid-body motion about fp** (GOLBEV, VV (Israel Progr. Transl.) ∋R-S analogues) (See also: LEIMANIS (bk)*, Springer)

2618. $\delta/@$ **Generalized resultants (anal. functions.)** (MR58 #2422)

2619. $\delta/\mathcal{A}$ **Graphs connectes with 'steepest descent proc.'** (MR 58 # 317)

2620. $\delta/\mathcal{A}/@$ **'Noncommutative analysis'** (Opr. Th. #57 (Birkhauser) 225–243) (Methods ∋ Maslov proc....)

2621. δ/A **Primes and Brownian Motion** (AMM 80 (1973) 1099–1115 / BILLINGSLEY)

2622. $\mathcal{A}/A$ **q-gram distance (string-matching)** (Seqs. II* R. Capocelli$^+$ eds. Springer, 1993 p300ff)

2623. $\mathcal{A}/A$ **Approximate tree-matching** (Ibid., p245, ff)

2624. δ **Converse of 'Divergence Thm.'** (AMM 71, 442–3, UNGAR, P)

2625. $\delta/\mathcal{A}$ **Complementary ineqs.** (Rend. Circ. Mat. Palermo XIII (1964) 291–328 JB DIAZ)

2626. δ/A **Probabilistic Metrics / Stability Stoch. Models** (SL RACHEV (Wiley, 1991) pp350, BLMS 27, 87)

2627. δ/A **Complete connections in random systems** (CT#96: IOSIFESCU, M. +(CUP, 1990))

2628. $\mathcal{A}/A$ **Approx. analysis of computer programs** (§10.6 of 'Progr. Flow Anal.' Muchnick +(eds) P-H, 1981)

2629. δ/A **Exp. ctty / formal differentiation of algorithms** (B PAIGE / JT SCHWARTZ 4th ACM Symp. 58–71)

2630. $\delta/\mathcal{A}$ **Boolean MS** (Thesis: CJ Penning (Delft) MR 23 #A179)

2631. δ/A **Use of differential geom. in Stats.** (NC GIRI 'Group Invariance in Stats' WorldSci., 1996*)

2632. δ/A **Use of cts. groups in multivariate calculation** (RH FARRELL (Springer, 1985)*)

2633. δ/A **Algebraic Fermi-Surface Curves** (BLMS 27, 413–)

2634. δ/A **Differential geom. techn. in Stats.** (BLMS 27, 619 (+ refs))

2635. $\delta/\mathcal{A}/A$ **Bounds on moduli for 2-component materials.** (PRSL A380, 305–331: GW MILTON +)

2636. $\delta/\mathcal{A}/A$ **Bounds on thermal conductivity (N-phase materials).** (PRSL A380, 333-348: PHAN THIEN)

2637. $\delta/\mathcal{A}/A$ **Strength of fabrics ('reflection')** (BLMS 26, 127–131 CRJ CLAPHAM)

2638. $\delta/@$ **Proof of 'theorems' from logical propositions** (MetaMaths of Algebra, A ROBINSON: N. Holland, 1951)

2639. $\delta/\mathcal{A}/A$ **'Uncertainty principles' for function reconstruction** (Duke MJ 42 (4) 661–706 (LOGAN))

2640. $\delta/\mathcal{A}/A$ **Min. Separation of stable / unstable matrices** (Contemp. math. 47 (1985) 465–477 C. Van LOAN)

2641. $\delta/\mathcal{A}/A$ **Approximation of curved spaces by quasi-polyhedra** (REGGE, R, Nuovo Cim. XIX (3), 1961)

2642. δ/A **Direct use of SAM in Robotics (?)** (DUFFY, J, Anal. Robot Manip. (Arnold, 1980)*)

2643. $\delta/\mathcal{A}/@$ **Ctd. Fractions whose components are 'op'ns'** (Duke MJ 26 (1959) 663–677, MACNERNEY)

2644. $\mathcal{A}/@$ **Approx Solutions to inconsistent lin. eq. systems** (e.g. PRATT, WK Digital Image Proc. (Wilet, 1978) p209ff)

2645. $\delta/\mathcal{A}/A$ **'Essential nature' of generalized harmonic anal.** (LEE, YW, statistical Th. Commun. (Wiley, 1960)*)

2646. $\delta/\mathcal{A}/A$ **Finite approximation of infinitely long logical formulae** (MR 34 #2464)

2647. $\delta/\mathcal{A}/A$ **'Re-topologization' (!): ?analogues of GF Th.** (NAGUMO Works * p407, Springer, 1993)

2648. δ/A **? Formalization of general / (in)organic reactions** (e.g. via SYKES' Guidebook...) [Also: 'CHEMTRAN' (MIT) / successors]

2649. $\delta/@/A$ **Alg. generalization of 'isomerism'** (SLANINA, Zd., 'Contemp. Th. Chem. Iso. Reidel 1986* p112ff)

2650. $\delta/\mathcal{A}/A$ **'Chemical distance / metric'** (Ibid. p115ff (e.g., 5.10))

2651. $\delta/\mathcal{A}/A$ **Graph-Th. / Topol. in Chem.** (Proc. (RB KING, ed.)* Elsevier, 1983)

2652. $\delta/\mathcal{A}$ **Detailed proc. numerical conformal maps** (See QAM 4/64 p46 (bk rev))

2653. $\delta/@/A$ **Comparison of 'unification' & 'uniformization'** (e.g. H/B of AI* Vol. 3 p89/CV bks) [Can these be representations of 'same' general procedure?!)

2654. $\mathcal{A}/A$ **Approx. complexity of functions** (BAMS 81 (1975) 112–3 / refs, RC BUCK)

2655. $\delta/\mathcal{A}/A$ **Rel. lengths of proofs in theories $\mathcal{J}, \mathcal{J}'$** (MOSTOWSKI, A, 'Foundational Studies' * Vol. 1, p29)

2656. $\delta/\mathcal{A}/A$ **Rev. of 'Classification Theory' (Skelah)** (BAMS 3/81p222–9) [Also, Wilf H. Survey, BLMS 1987*]

2657. $\delta/\mathcal{A}/A$ **Predictive decompositions of Time-Series** (Geophysics XXXII 1967, 418–484, EA Robinson / bk*)

2658. $\delta/\mathcal{A}/A$ **Discrete 'inverse methods (elast. waves)** (Ibid. 45 (1980) 213–233, BERRYMAN, JG)

2659. $\delta/\mathcal{A}/A$ **Stability results (generalized matrix inv.)** (Optimization Geodetic NW F. HALMOS (ed.) 307–321, B BAROTHY)

2660. $\delta/@$ **Convolutional 'square roots'** (J. Anal. Math. 10 (19/1963), 363–379, KOREVAAR, J MR 26 #6700)

2661. $\delta/\mathcal{A}/@$ **'Calcul Diff. Espaces Vect. Quasi-topol.'** (MR 54 #11375, M SIMMONET)

2662. $\delta/\mathcal{A}/A$ **Parallel approximation procedures for optimization** (CMV-CS-84-155* J PETERS / L RUDOLPH)

2663. $\delta/@/A$ **Unsolvable decision problem $\to$ undecidable word problem** (See MR 20 ##5230/1, W BOONE) [via intro. of 'formal inverses']

2664. $= 2593.$ $(\mathcal{A}/A)$

2665. $\delta/@$ **Inverse (eigenvalue) problems** (LNM #228*, OSBOURNE, MR: pp155–168)

2666. $\delta/@$ **'Compositional primality'** (MR 53 #11030, F GROSS)

2667. $\delta/\mathcal{A}/A$ **The inverse of a queueing process** (MR 40 #2179, L TAKÁKS)

2668. $\delta/\mathcal{A}/A$ **Econ. location / measure th.** (AM FADEN 'Econ. of Space & Time' (Iowa St. UP) pp703, 1977, BAMS 5/79 pp360ff (rev.))

2669. δ/A **Combinatorics 'via' Graph Theory** (JE GRAVER / ME WATKINS, pp351 (Springer, 1977)) [Unified scheme: ?BLMS rev.]

2670. $\delta/\mathcal{A}/A$ **Probabilistic MS / Hysteresis Systems** (Comm. Math. Phys 20 (1971) 205–219)

2671. $\delta/\mathcal{A}/A$ **Stability (anal./alg./replication) in Multivariate Analysis** (See: 'Nonlinear MV Analysis', by A GIFI*, Wiley, 1990)

2672. $\delta/\mathcal{A}/A$ **Functional Analysis / Valuation Theory** (MR 50 #14142; 45 #4101, NARICI⁺, AP, 1971) [Also: HAZEWINKEL, M, bk(s) ...]

2673. $\delta/@/\mathcal{A}$ **Extension of topological structures** (See Proc. FLACHSMEYER (ed.)* VEB 1969)

2674. $\delta/\mathcal{A}/@$ **Extension of anal. objects** (Yan, Y-T: (Dekker, 1974)*)

2675. $\delta/\mathcal{A}/A$ **Stability of Stochastic Models** (RACHEV, ST (bk), Wiley 1991 BLMS #124, pp87–89 (rev.))

2676. $\delta/\mathcal{A}/@$ **Product-Integration** (BAMS 3/82, 230ff (rev.) of DOLLARD / FRIEDMAN)

2677. $\delta/\mathcal{A}$ **Evaluation of** $\int_{\mathbb{R}} \exp\{(\bar{z}, HZ)\}dxdy$ (Duke MJ 26 (3) 1959 485 (Bezlman))

2678. = 2643. ($\delta/\mathcal{A}/@$)

2679. $\delta/\mathcal{A}/@$ **Constructive proc. / probabilistic MS** (Fund. Math. 67 (1970) 115–124; NISHIURA, E)

2680. $\mathcal{A}/A$ **Hausdorff-measures... of size for sets of $\mathcal{L}$-measure O** (e.g. CA ROGERS, 'H'sdorff Meas.' (CUP)*)

2681. $\delta/\mathcal{A}/A$ **Ultra-MS / Fractals / Chem. Reactions** (PRSL A423, 189–200, M BLUMEN / KÖHLER) [PRSL A423 pp1–200 'Fractals in Nat. Sci.]

2682. $\delta/@$ **Groups of infinite matrices** (?refs.)

2683. $\delta/\mathcal{A}/A$ **Lin. treatment of nonlinear problems** (See 'Controls & Optimization', JE RUBIO, Manchester UP) [reduction to ∞-dim. LP via Radon transforms.]

2684. $\delta/\mathcal{A}/A$ **Thresholds for random graphs** (Random Gr. (Bóllobás, AP185) / Gr. Evolution, EM PALMER (Wiley, 1985))

2685. $\delta/\mathcal{A}/A$ **(Almost) (non) smooth optimization** (Convex Anal. / Minimization Algorithms, Springer 1993* Vols 1, II, HIRIART-URRUTY$^+$)

2686. $\delta/@$ **Representation of $f(x)_n$ 'via' $\{g_k(x_l)\}$** (Soviet Math. Dokl. 8 (1967) 1550ff*)

2687. $\delta/\mathcal{A}/A$ **Probabilistic vs. deterministic structural stability** ('Probabilistic Methods Str. Eng.' Augusti E^{++} C&H 1984, 438$^+$)

2688. δ/A **Deformation Waves (structural anal.)** (Distribution of Deformation', V KLOUČEK (Artia, 1955))

2689. $\delta/\mathcal{A}$ **Iterative Solution W-H $\int$-eqs.** (QAM XX 1/63: $(T^2WU)^2$)

2690. δ/A **Noncommuting random products** (MR 29 #648: H FIRSTENBERG)

2691. $\delta/\mathcal{A}/A$ **Approximation of Hopf bifurcation** (Numer. Math 39 (1982) 15–37: C BERNARDI)

2692. $\delta/\mathcal{A}/A$ **Statistical proc. 'as' optimization problems** ('Math. Progr. in Stats', ARTHANARI / DODGE WILEY v.1981)

2693. δ/A **Combinatorial aspects of Schubert Calculus** (LNM #574, 217–251: RP STANLEY)

2694. δ/A **Calculus of data refinement** (Progrm. / Math. Method (D. BROY, ed.) Springer 1992, 213–244 (D. GRIES))

2695. $\delta/\mathcal{A}/A$ **'Freq. spectra' of TS** (Trans. Moscow MS (1981)* 163–200)

2696. $\mathcal{A}/A$ **Approximation problems in anal./synthesis** (Anal. / Synth. Lin. Time-Var. Syst., AR STUBBERUD (bk)*)

2697. $\mathcal{A}/A$ **MS of evolutionary distances** (J. Algor. 1 (1980) 359–373 (Sellers, D / SANKOFF, D, (ed.)) 'Time Warps...' bk* AW 1983)

2698. $\delta/@$ **Path algebras** (e.g., CARRÉ, B, Graphs / NW: OUP, 1978*)

2699. $\delta/\mathcal{A}/A$ **Theorems as perturbations of knowledge states:** → **'response functions', etc!** (To be developed!)

2700. $\delta/\mathcal{A}$ **Construction of functions with arbitrary spec. singularities** (HOBSON, Th. R / Functions, Vol II, Ch. VI)

2701. $\mathcal{A}/A$ **Approx. e-functions / e-v** (J. Optimal Soc. America 66 (1976) 525–9)

2702. $\delta/@/\mathcal{A}$ **Calculus in quasi-TS** (MR 54 #11375 / LNM ##417/469/540/...)

2703. $\delta/@$ **Solution of 'invariate subspace' problem (BS)** (BLMS 16 (1984)* 337–401, CJ READ)

2704. $\delta/@$ **Anisotropic function spaces** (Anal. Mathematica 10 (1984) 57–77/77–96; H TRIEBEL) ?∗

2705. δ/A **Approximation of convex functions by rat^2 functions** (Anal. Mathematica 10 (1984) 15–21; HATAMOV, A)

2706. $\delta/\mathcal{A}/A$ **Wiener's Tauberian theorem 'as' propagation** (?e.g. H REITER bk / OUP∗)

2707. $\delta/\mathcal{A}/A$ **Simplicial approximation in statistical circuit design** (Circ. Th. / Design BOITE$^+$ (eds) 15–24)

2708. δ/A **Matroids in Circ. Th.** (Ibid., 164-175)

2709. δ **Inf. Ctd. Fraction of Oper.-valued anal. functions** (Ibid. 186–189)

2710. $\delta/@/A$ **Hankel norms in model reduction algorithms** (Ibid. 205–212/222–233)

2711. $\delta/\mathcal{A}/A$ **Dynamics 'as' limitation proc. for microstructure** (ARMA 133 (1996) 199–247, G FRIESECKE / JB McL.)

2712. $\mathcal{A}/A$ **Helinger (& other) distances (Prob. Th.)** (REISS, RD: 'Point Proc.', Springer 1993∗)

2713. $\delta/@$ **'Complete' M-L decompositions of Mero. functions** (QAM XXVII / 1969, 185–192, CURME / HILLE)

2714. $\delta/\mathcal{A}/A$ **Solvability / approximability of Nonlin. NW** (e.g. DOLEZAL, V, Monotone Operation / Applications: BAMS 3/80 369 rev.

1979, pp 174, bks*) [∋ Abstract NW / general criteria: cf Abstract System Theory]

2715. $\delta/\mathcal{A}/@$ **Properties 'true pp' / hard or impossible to realize** ('Some Random Series of Functions' (bk)* JP KAHANE (HEATH, 1968)) [This involves 'stochastic ME', ...]

2716. $\delta/\mathcal{A}/@$ **Limits in (co)homological algebra** (e.g. STROOKER, JR*, 'Intro t Categories, Homological Algebra & Sheaf Cohomology' CUP, 1978, esp. Ch. 1) [see also BAMS 11/79* 919–927]

2717. $\delta/\mathcal{A}/A$ **Chemical-reaction waves** (J. Chem. Phys. 60 (1974) 5090–5107, OTROLEVA / ROSS #90 in JCP list)

2718. $\delta/\mathcal{A}/A$ **Hydrodynamic analogue of QM** (J. Chem. Phys. 60 (1974) 2762–6, A ASKAR$^+$ #210 in JCP list)

2719. $\delta/\mathcal{A}$ **Local non-convexity** (Israel JM 10 (1971) 196–209, MD GUAY / DC KAY) [Also: SR LAY, 'Convex Sets / Applications' (Wiley, 1982)* Ch. 7: 'Local$^+$ ⇒ Global']

2720. $\delta/\mathcal{A}$ **Surface area / integration VIA inscribed polyhedra** (Math. Naehr 99, 105–114 +refs., TORALBALLA, LV/LC*)

2721. $\delta/@$ **Matrix-W-H-decomposition** (PRSL CA393*, 185–192, DS JONES) [M=: M_R+M_L / M =: M^+M^-: systems of W-H $\int$-eqs.]

2722. $\delta/\mathcal{A}/A$ **Approximative formulation of GR$^-$** (Synge, JL, 'GR' N-Holland 1960*) [∋ PS representations in geodesic separation]

2723. $\delta/@/\mathcal{A}$ **Continuous induction** (Khilim, G, 'Qual. Methods in the Many-Body Problem': G&B 1961, Ch. 3*)

2724. $\delta/\mathcal{A}/A$ **Cellular automata for partially recursive functions** (JACM 18 (3) 1971 339–353: AR SMITH III)

2725. $\mathcal{A}/@/A$ **Duality: abstract sb.-hm. functions / Jensen Measures** (LMS Lec. N. #32 GAMELIN, TW)

2726. $\delta/\mathcal{A}/A$ **p-adic interpretation of various analytical operations** (LMS Lec. N #46, N KOBLITZ)

2727. $\mathcal{A}/A$ **Distance (prime ideal, module)** (BLMS 3/79, P DUTTON)

2728. $\delta/\mathcal{A}/@$ **Inverse measures ($\mu t(dt)$) of $\mu(dt)$** (Adv. Applied Math. 18 (1997) / MPCPS 123 (1998) Mandelbrot, B / RIEDI) $[\models (\mu t)t = \mu]$

2729. $\delta/\mathcal{A}/A$ **(Quasi-)Riemannian Colour Metrics** ('Digital Image Proc.', WK PRATT (Wiley, 1978)* p.170→) [Also: brightness-(?) metrics]

2730. $\delta/\mathcal{A}/A$ **Analogue of Riemann Mapping Theorem for Lorentz Metrics** (PRSL A401, 1985, 119–13, RS KULKARNI)

2731. $\delta/\mathcal{A}/A$ **Information-Content / 'Set-size'** (O WATANABEE, ed., Kolmogorov-Complexity and Computational Complexity, Springer, 1992, 23–42)

2732. $\delta/\mathcal{A}/A$ **Probabilistic proof of real LT Inversion** (MPCPS 125 (1999) 139-149, JA ADELL / C SANGÜESA)

2733. $\delta/@$ **Iterated Path-Integrals** (BAMS 83 (1977) 831–879: KT CHEN)

2734. $\delta/@/A$ **Nonstandard Analysis VIA 'Internal Set Theory'** (BAMS 83, 1165–98: E NELSON)

2735. $\delta/@$ **Analytic (holomorphic) curves** (BAMS 83, 553–568: B SHIFFMAN)

2736. δ/A **CV techniques in Algebraic Geometry** (BAMS 1 (1979) 595–626, PA GRIFFITHS)

2737. δ/A **Intersection of Quadrics** (Russian Math Surveys 30, (6) 1975, 51–105, AN TYURIN)

2738. δ/A **Real forms of complex algebraic curves** (Trans. Moscow Math. Soc. 51, (1988) 1–51: SM NATANZON)

2739. $\delta/\mathcal{A}/@$ **Almost-similarity of operators** (Ibid. 36 (1978) 57–82, GV ROZENBLUJM)

2740. $\delta/\mathcal{A}/A$ **Criteria for Quasi-Powed-Bases in** $\mathcal{A}(D)$ (Russian Math. Surveys 30 (1975) 101–146, I IBRAGIMOV / NAGNIBIDA)

2741. $\delta/\mathcal{A}/A$ **Generalized Covering Theorems** (WJ Trjitzinsky: 'La Regularité Moyenne dans La Théorie Métrique')

2742. δ/A **Riemannian structures on {T-D states}** (Rend. Mat. Acc. Lincei S.9, V.3, 45–50 (1992) NAPOLITANO, LG / ALBANESE, C)

2743. δ/@ **Normed Division Domains** (AMM 88 (1981) 681–6 (SW GOLOMB)) [∋ generalizations of 'division' operations: e.g. on GRAPHS (? ∋ Σ -diagrams)]

2744. δ/A **Accoustic signatures of cracks** (PRSL A436 (1992) 251–265, DA REBINSKY / JG HARRIS)

2745. δ/@ **Dimension of (orderings/) graphs** ('Nonserial Dynamic Programming', Bertele, U / Brioschi, F (AP, 1972)*)

2746. δ/@/A **Quantization as (i) mapping, (ii) deformation of poly. algebras** (Studies Math. Phys. #4, AO Barut (ed.), J NIEDERLE: pp83–97) [Also other papers in this volume]

2747. $\mathcal{A}$/A **Extreme / exposed points in BS...** (BLMS 28* (1996) #130, 51–58, FONF, VP)

2748. δ/@/A **Motion of quasi-particles in elastic solids** (Quasi-particle Th. of Defects in Solids (DI PUSHKAROV), World Scientific, 1992*) [See also: 'Th. of Molecular Exitons' (AS DAVYDOV, McGr-H, 1962)*]

2749. δ/@/A **Stochastic integration of differential forms over random k-chains** (LNM #866, JM BISMUT) [See BLMS 14 (1982) 450–1 (rev.) (Mécanique Aliétoire)]

2750. δ/$\mathcal{A}$/A **Differential Geometry & Stats.'** ((1) 'Multivariate Calculation', by RH FARRELL (Springer, 1985), (2) 'Differential Geometry & Stats.', by MK MURRAY / JW RICE (C&H, 1993))

2751. $\delta/@/A$ **Inaccessibility of stable states from META stable states** (T-D / S-M / Q-M bks)

2752. $\delta/\mathcal{A}/A$ **Transmission-LIMITATIONS of MONOCHROMATIC / PLANE Waves** (E-M bks)

2753. $\delta/@/A$ **Abstract Functional Continuation (∋ DAC)** (Ann. Mat. Pura Appl. (4) 49 (1960) FICHERA, G (MR 22 #8495))

2754. $\delta/@/A$ **Topol. Generalizations of 'Fundamental Thm. of Algebra'** (See, MR 23 #A2195, REICHBACH, M)

2755. $\delta/\mathcal{A}/A$ **Rates of convergence in Central Limit Thm.** (Pitman Res. Notes #62 (1982) P HALL)

2756. $\mathcal{A}/@$ **Bounds on functions of matrices** (AMM 74 (8) 920–926)

2757. $\delta/@$ **Statistical Circuit Design / Simplicial approximation** (∋ pp15–24 of 'Circuit Theory & Design', R BOITE (eds), N-H, 1981*)

2758. $\delta/\mathcal{A}/A$ **Inverse Prob. for ODE: Dynamical Solutions** (Yu S OSIPOV / VA KRYASHINSKII, pp625 (G&B, 1995) CONSTRUCTIVE)

2759. δ/A **Formulas for Stress / Strain / Structural Matrices** (WD PILKEY, Wiley 1994, pp1458)

2760. $\delta/\mathcal{A}/@$ **Generalization of 'Covering Theorems' (∋ Vitali's)** (See AMM 74 (1967) 1281. rev. of bk., WJ TRJITZINSKY (French))

2761. δ/@ **Open Map / Cl.-Graph Theorems are $\approx$ equivalent** (See MR 22 #5964 (V PTAK) and MR 31 #2589 (bk: T HUSSAIN)*)

2762. $\delta/\mathcal{A}$/@ **Stable Banach Algebras** (See MR 56 #16375 / LNM #609, 109–114 (BE JOHNSON)) [$\exists \varepsilon > 0 : \pi, \rho$ 'multiplications' on $A \wedge \|\pi - \rho\|_{L^2(A)} < \varepsilon \Rightarrow \exists T, T^{-1} \in L^1(A) : \forall a, b \in A, T\rho(a,b) = \pi(Ta, Tb)$]

2763. δ/@/A **Rev. of 'Assigning Probabilities...'** (Scott / Krauss) [Synthese 17 (4) (1967) 456–9 (FENSTAD)*)

2764. $\delta/\mathcal{A}/A$ **Approx. / almost-sure solution of 'hard combinatorial problems'** (MR 53 #7121 (RM KARP)) [Problems $\ni$ solution time/effort $\approx r \times$ (optimal feasible solution time/effort)]

2765. $\delta/\mathcal{A}/A$ **Stability of Dirichlet Problem** (See LANDKOF, 'Foundations of Modern Potential Th.' (Springer, 1972)*)

2766. δ/@$\mathcal{A}$ **Ring-theoretic schemes in Control Theory** (Poly. Methods... (KJ HUNT, ed.), IEE Ser. #49, 1993*)

2767. $\delta/\mathcal{A}/A$ **Operational-Th., Analytic Functions, Matrices, Elec. Engineering** (CBMS #68, AMS, 1987, pp134, JW HELTON) [see: BLMS 11/90, p614-6]

2768. $\delta/\mathcal{A}/A$ **Cts.-time random walks** (e.g. PRSL A423, 189–200 (A BLUMEN$^+$) (for chemical reactions in 'fractal media'))

2769. δ/@$\mathcal{A}$ **Tensor operations / 'strings' in statistics** (PRSL A406, 127–137 +refs, BARNDORFF-NIELSEN, OE)

2770. $\delta/@$ **Fredholm Theory in BS*** (CUP, 1986, pp293, AF RUSTON)

2771. $\delta/@\mathcal{A}$ **TAXONS $\to$ optimal set-partition(s)** (See VAPNIK, V, Estimation of Dependencies, Springer, 1982*, App. to Ch. 10 & refs. there)

2772. $\mathcal{A}/A$ **Metric properties for SMS** (JLMS 39 (1964) 117–128 (Rhodes) 129–130 (Kingman)) [Metrics / topol. / convexity...]

2773. $\delta/@/A$ **Discrete models of QM** (J. Math. Phys 27 (1986) 1782–90, SP GUDDER)

2774. $\delta/\mathcal{A}/A$ **S-MS / Hysteresis Systems** (MR 43 #5853, Erber, T_{++}, Comm. M. Phys. 20 (1971) 205–219)

2775. $\mathcal{A}/@$ **Approximation of Σ RV by ∞-divisible distributions** (Dokl. 24 (1981) 382–5 ZAICEV, AY) [The Lévy metric is used]

2776. $\delta/@$ **Converse of Max. Mod. Thms.** (see MR58 #6218)

2777. $\delta/\mathcal{A}/A$ **Baire-pp Stoch. DE have unique strong solutions** (Ann. Prob. 14 (1986) 653–662, AJ HEUNIS)

2778. $\delta/A/@$ **Discrete analogues of singular sequences*** (Trans. Moscow MS, 1984, 43–67, Belyankov / Ryabenskii)

2779. $A/@$ **Stochastic bounds:** $\forall \varepsilon > 0, P\{|X_n| > a\} < \varepsilon$ (Feller, Vol II, p247)

2780. A **Weakly (/locally) semi-algebraic spaces** (LNM ##1173 / 1367*) ($\ni$ diverse approach criteria...)

2781. $\mathcal{A}/@/A$ **Boolean Methods in Approximation** (Pitman Res. N #230, DELVOS / SCHEMP, 1989)

2782. $\mathcal{A}/A$ **Approximation in Nevanlima Theory** (LNM #1433*, 1990, LANG S / CHERRY, W)

2783. $\delta/\mathcal{A}/@$ **Generalizations of Contraction maps** (Dokl. 15 (1974) #2*, 673–6, AD GORBUNOV)

2784. $\delta/@$ **Multiplicative analogue of Laurent's (CV) Theorem** (PLMS (1956) 59–69, P MASANI) [Also Ibid. 43–58: Rational approximation of op.-valued functions]

2785. $\delta/@/A$ **Generalized Lévy / Khinchin representation for T-Groups** (Th. Prob. Appl. XI (1) (1966) 1–45, VV SAZONOV+, VN TUTUBALIN)

2786. $\mathcal{A}/@$ **Contractions for 'almost-metrics'** (MPCPS 119 (1995) 31–55, SP EVESON / RD NUSSBAUM)

2787. $\delta/@$ **Analytic functions in LC Algebras** (PLMS (3) XVI 321–341 (1966) RJ ELLIOTT)

2788. $\delta/\mathcal{A}/A$ **Local Theory of BS** (LNM #1200 / CUP Tract (Pisier, G))

2789. $\delta/\mathcal{A}/@$ **Predominantly contractive maps** (JLMS No. 149 (1/63) 81–86, M EDELSTEIN)

2790. $\delta/@/A$ **Method of Images (E-M) / Riemann Surfaces** (Sommerfeld (¿PDE') / Carslaw (obit[+]))

2791. δ/A **Coordinate-free representation of statistical strings'** (PRSL A415, 445–452 (1988) MK MURRAY)

2792. $\delta/@/A$ **Generalized P-W Theorem** ($\Rightarrow$ **E-of-W dg. Theorem**) (MPCPS 105 (1989) 285–297, A PULTR)

2793. $\delta/@/\mathcal{A}$ **Metric Structures over frames / locales** (MPCPS 105 (1989) 389–395, TG GENCHEV)

2794. $\delta/@/A$ **'Pressure' of cts. function of topol. dyn. system** (Walters, P, Intro Ergodic Th. (Springer, 1982)) [See also: JLMS (2) 47 (1993) 142–156 (F HOFBAUER))

2795. $\delta/@/A$ **Stochastic analogues of QM processes** (PCPS 64 (1968) 1061–1070, JG GILSON)

2796. δ/A **Continuum- vs. Lattice-based Markov Chains** (PCPS 61 (1965) 173–190 (J KEILSON / DMG WISHART))

2797. δ/A **Analogues of phase diagrams, etc.** (cf. RICCI, JE, 'The Phase Rule...'*, Dover, 1966)

2798. $\delta/\mathcal{A}/@$ **Jentzsch-type theorems for Dirichlet series** (See MR 26 #301; 30 #234, POPOV, AF / FOMENKO, SV)

2799. δ/A **Noncentral χ^2-distribution via contour integration** (Ann. Math. Stat. 33 (1962), 796–800, McNOLTY, F)

2800. $\delta/@$ **Functions of finite Markov chains** (Ann. Math. Stat. 34 (1963) (i) 1022–1032, (ii) 1033–1041, DNARMADHIKARI, SW)

2801. δ/A **Bundel-maps representation of RHEOMETRIC INVARIANCE** (PRSL A372, 169–200, Sec. 3, B CARTER)

2802. $\delta/@/A$ **Logic / Differential-Geometric Procedures** (AJ WILKIE (Ox.))

2803. δ/A **Undecidability theorems for Progress Schemata** (SIAM J. Comput. 1 (1) (1972) 119–129, RE MILLER *)

2804. $\delta/\mathcal{A}/@$ **Unified Representations of Quadrature Rules** (SIAM J. Numer. Anal. 9 (4) (1972) 573–602, JD DONALDSON / D ELLIOTT)

2805. $\delta/\mathcal{A}/@$ **General representations of remainders in numerical methods** (SIAM J. Numer. Anal. 9 (3) (1972) 476–492, B CHARTRES / R STEPLEMAN)

2806. $\delta/\mathcal{A}/A$ **Homological perturbation theory** (21st Nordic Congr. Mathematicians 6/92, L LAMBE+ refs.) [Also: PAMS 112 (1991) 881–892, BARNES, D / LAMBE)

2807. $\delta/\mathcal{A}/A$ **'Almost all' finite games have odd no. of equilibrium points** (Int. J. Game Th. 2 (1973) 235–250, MR58 26059, HARSANYI, J) [Original proof: MR51 #7642, RB WILSON, SIAM J. Appl. M. 21 (1971) 80–87]

2808. $\delta/\mathcal{A}/A$ **Word-problem formulation of bifurcation** (MPCPS 108 (1990) 127–151 J DEVLIN)

2809. $\delta/@/A$ **Analysis on fractals: effective resistances** (MPCPS 115 (1994) 291–303, J KIGAMI)

2810. $\delta/\mathcal{A}/A$ **A Matrix Approach for Proving Ineqs.** (A FERSCHA ?ref. (SAM) Proc.)

2811. $\delta/\mathcal{A}$ **Distribution of EXTREME values** ('Statistics of Extremes'*, Columbia UP (NY), EJ GUMBEL)

2812. $\delta/@$ **Finitely-additive measures (charges)** (KPS Bhaksara-Rao et al., AP (1983)) [See: BLMS 16, 431 (1984)]

2813. $\delta/\mathcal{A}/A$ **'Algorithmic Algebraic Number Theory'** (M POHST / H ZASSENHAUS, CUP ($1988^{(+)}$))

2814. $\delta/@$ **Integral ineqs.** $\ni f$ **and** f' (AMM 78 (7) 705–741 (1971), PR BEESACK)

2815. $\delta/\mathcal{A}/@$ **Metrics / analytic functions in Alg. Number Fields** (e.g. BOREVICH / SHAFOREVICH Number-Th (AP, 1966))

2816. $\delta/\mathcal{A}/A$ **Belevich: Classical N-W Th.** (MR 39 #5243)

2817. $\delta/\mathcal{A}$ **Elements of Computer Algebra** (AG AKRITAS: pp426, Wiley, 1989)

2818. $\delta/@$ **Constructive duality in Orlicz Spaces** (MPCPS 89 (1981) 49–69, DL JOLINS / CG GIBSON)

2819. $\delta/@$ **'Universal Matrices'** (ibid. 85 (1979) 193–198: H BURKILL)

2820. $\delta/@$ **Formulae for $\partial_{a_{ij}}^m r(A)$ [Spectral radius of A]** (ibid. 83 (1978), 183–190, JE COHEN)

2821. $\delta/\mathcal{A}/A$ **Hopf bifurcation for BLOCK-DIAGRAM representations** (ibid. 82 (1977) 453–467, DJ ALLWRIGHT)

2822. $\delta/\mathcal{A}$ **Algorithm for (partial) poly. factorization** (ibid. 82 (1977) 427–437, MR FARMER / G LOIZOU)

2823. $\delta/@/\mathcal{A}$ **Asymp. properties of alg. functions of n vars.** (Russ. Math. Surveys 16 (1961), 93–119, EA GORIN)

2824. $\delta/\mathcal{A}/@$ **Choqunet boundaries in Kantorovich spaces** (ibid. 30 (1975) 115–155, SS KUTATELADZE)

2825. $\delta/\mathcal{A}/@$ **Direct / Converse Theorem (Approximation Theory)** (ibid. 23 (1968) 115–177, NP KUPTSOV)

2826. $\delta/\mathcal{A}/@$ **Limit distributions for iterated poly.** (MPCPS 96 (1984) 237–253, P FLAJOLET / AM ODLYZKO)

2827. $\delta/\mathcal{A}/A$ **Operator analogues of CV Theorems** (M. Zeit 160 (1978) 275–290: Ky Fan) [∋...Schwarz Lemma / Pick Theorem / Vitali Theorem (all fr $A : \|A\| < 1$)]

2828. $\delta/\mathcal{A}/A$ **Sensitivity of Nonlinear Progress to Parametric Changes** (e.g. McCormick, GP, 'Nonlinear Progress.'*, Wiley, 1983, §11.3)

2829. $\delta/\mathcal{A}/A$ **Weak stochastic coupling (of external structures)** (PRSLA395, 141–151, P WHITTLE (1984))

2830. $\delta/\mathcal{A}/@$ **Semigroup decomposition of automata** (e.g. MRI Symp. XII (Polytechnic Pr.), 1962, Math. Th. Aut. pp341–384, KROHN / RHODES) [Also: Proc. M ARBIB (ed.) AP, 1968]

2831. $\delta/@/A$ **'A Mathematical Theory of Systems Engineering'** (AW WYMORE, pp353, Wiley, 1967 (TR: AMM 75, 434))

2832. $\delta/\mathcal{A}/A$ **Average asymmetry of m Polymers** (PRSLA432 (1991) 495–533, KM JANSONS) ($\ni$ Grom. prob. / BM-measures...)

2833. $\delta/@/A$ **Hyper-singular $\int$-equations** (e.g. PRSL A432 (1991) 301–320, PA MARTIN)

2834. $\delta/\mathcal{A}/A$ **Combined schemes for statistics / control** (P BAUER, et al. (eds), Reidel, 1987) [P MANDL: pp155–168 in 'Mathematical stats. / probability: Vol B']

2835. $\mathcal{A}/@$ **Approximation of curved spaces by quasi-polyhedra*** (Nuov. Cim. XIX (3) 558–571; T REGGE)

2836. δ/A **Physical interpretation of Reimann-Christoffel Tensor*** (Tensor (NS) 4 (1955) 150–172; G KRON) [Damping / torques in osc. transmission systems]

2837. $\delta/@$ **Constructive W-H Factorization** (I GOLIBERG + (eds), Birkhäuser, 1986)

2838. δ/A **Benzecri, JP, 'L'analyses des données': 1. 'La Taxonomie'; 2. 'L'analyse des correspondences'** (Dunod, 1976)

2839. $\delta/@/A$ **Dynamics of Mechanical / Electrical / Hydraulic / Pnmt. Systems** (OGATA, K, pp596, P-H, 1978)

2840. $\delta/@$ **Quasi-constructive pf. of Imp. FT** (See MR 21 #705; 19 #176, HART, WL / MOTZKIN, TS)

2841. δ/A **Applications of Symb. Computation** (JC HOWARD, Tech. P., 1980, pp413)

2842. $\delta/@/\mathcal{A}$ **n-variable spectral-th.** (Functional An. (Symp.) AP (1970) JL TAYLOR) [See MR42 #6623 (long-), 65–143]

2843. $\delta/@$ **Topol. FP Th. / n-val. maps** (Russ MS 35 (1980) / Yu G Borisovich ++)

2844. $\delta/\mathcal{A}/A$ **Distance from a system fo {uncontrollable systems}** (LNCIS #58, Springer, 1984, 302–314, R EISING*)

2845. $\delta/\mathcal{A}/@$ **Ctd. fractions over non-commutative algebras** (LNCIS #58*, 279–292, A DRAUX)

2846. $\delta/\mathcal{A}/A$ **Systems with noncommensurate time-delays** (LNCIS #58*, 520–540, EW KAMEN ++)

2847. $\delta/\mathcal{A}/A$ **Some Algebraic Varieties in Syst. Th.** (LNCIS #58*, 541–9, D KANEVSKY)

2848. $\delta/\mathcal{A}/A$ **Spectra of finite Toeplitz Matrices** (LNCIS #58*, 194–213, P DELSARTE +)

2849. $\delta/@/A$ **Simplicial decomposition of (∞) graphs** (R Diestel: 'Gr. Decomposition...', OUP 1990) [See BLMS 24 (1992) 90–2]

2850. δ/A **Chemical Models...** ('Chem. through Models', CUP, 1978, CJ / KE / CW SUCKLING)

2851. $\mathcal{A}/@/A$ **Approximation of HS Operators** (Pitman Res Notes #72, 102 (1982, 1984), DA HERRERO (72); C APOSTOL (43) (102) 1982, 1984)

2852. $\delta/@/A$ **Regge calculus in QFT** (PRSL A383, 359–377 (1982) NP WARNER)

2853. δ/A **Folded developables** (Thin-sheet structures) (PRSL A383, 191–205, 1982, JP / JL DUNCAN)

2854. $\delta/\mathcal{A}/@$ **Generalized polynomial expansions / representations** (A ROBERT, 'Systemes de Polynomes', Q's Univ., 1973 (Ont.))

2855. $\delta/\mathcal{A}/A$ **Top. proc. for group presentations** (DE COHEN, QMC Notes (1975) / 1989 CUP)

2856. $\delta/@/A$ **Centralized curves (cf GFs)** (LC YOUNG, 'Lecs. of Calc. Vor.', Saunders, 1969*)

2857. $\delta/@$ **Noncommuting random products** (TAMS (1963) 377–428, MR: 29 #648, H FURSTENBERG)

2858. $\delta/@/A$ **Ohm's Law for dislocation distributions** (MR 22 ##5204–6, EF HOLANDER)

2859. $\delta/\mathcal{A}/A$ **'Scanning approximation'** of $\int \underset{n}{\ldots} \int$ by $\int_\zeta$ (MR 21 (1960) #784 ∋ m refs.)

2860. $\delta/\mathcal{A}/A$ **Stability of group-representations as direct / inverse limits** (MR 21 #872, HB GRIFFITHS)

2861. $\delta/@$ **Inequiv. definitions of 'simple connectedness** (AMM 74 (1967) 1117–20, R JOHNSONBAUGH)

2862. $\delta/\mathcal{A}/A$ **Random nonlin. fp. theorems / $\int$-eqs.** (Math. Syst. Th. 11 (1977) 77–84, ACH LEE / W PAAGETT)

2863. $\delta/\mathcal{A}/@$ **Stone-W / Boolean Approximation / Interpolation** (Bk: FJ DELVOS / W SCHEMPP, Longmans 1989, pp 168) [Also: Acta Applic. Math. 23 (1991) 297 (rev.)]

2864. $\delta/\mathcal{A}/A$ **Probabilistic interpretation of FS-Agce** (Acta Applic. Math 23 (1991) 261–273, C BOLDRIGHINI (+2))

2865. $\delta/@$ **'Nonlinear Superposition'** (LNCIS #41* M Gössel)

2866. $\delta/@$ **'Near-Clty. (etc.)** (AMM 74 (8) p1036 [A IRUNDAYANATHAN!])

2867. $\delta/@/A$ **Random nonlinear contractions** (Math. Syst. Th. 11 (1977) 77–84, ACH LEE / W PAGETT) [Also: BAMS 82 (1976) 641–657 (Bharucha-Reid)]

2868. $\delta/\mathcal{A}/@$ **Boolean interpolation / approximation** (Pitman Res. Notes (1989) DELVOS / SCHEMPP)

2869. $\delta/\mathcal{A}/A$ pp 'stochastic / regular DE' have unique solutions (Ann. Prob. 14 (1986), 653–662: HEUNIS, A) [≡ #1144]

2870. $\delta/\mathcal{A}/A$ **Control-th. reconstructability** (A JOHNSON, 'Process Control...', Peregrinus 1985*)

2871. $\delta/\mathcal{A}/A$ **Topol. characterization of chemical N-W** (SIAM J. Appl. Math. 15 (1967) 13–68: Peter SELLERS / AG LARMAN / CA ROGERS)

2872. $\delta/\mathcal{A}/@$ **Normability of metrizable sets** (BLMS 5 (1968) 39–48)

2873. $\delta/\mathcal{A}/A$ **Iso-triviality of (m. tr.) S-groups** (MR 35 (1968) #6690 ∋ refs.) (≡ splittability after removal of finite étale cover of S)

2874. $\mathcal{A}/\delta$ $x : Ax = b, A \geq 0$ (SIAM J. Alg. Disc. Meth. 3 (1982) Y EGAWA / SK JAIN, 197–213)

2875. $\delta/\mathcal{A}$ **Geom. of Mero. Functions** (Mat. Sb. 42 (1982) 155–196, GA BARSEGJAN)

2876. $\delta/@$ **Dimensional deformation of $\int$-ble systems** (Inverse problems 5 (1989) 67–86, PM SANTINI / Also: Stringer symposia)

2877. $\delta/\mathcal{A}$ $\int$ **and differential representations of solutions of SLA Eq.*** (VL GIRKO: 'Statistical Anal. of Observations...', KLUVER, 1995)

2878. δ $\int$**-representation of** $R_o^{-l-u} \equiv [\rho^2 + \rho_0^2 - 2\rho\rho_0\cos(\phi - \phi_0) + z^2]^{\frac{-l-u}{2}}$ (VI FABRiKANT: Potential Th. & Mech. Bk.)

2879. $\delta/\mathcal{A}/A$ **Analogues: #Th / Dynamical Systems** (LMS LN #134 MM DODSON / JAG VICKERS / BLMS 22 (1990) 612–)

2880. $\delta/@/\mathcal{A}$ **Almost convergence of double-seqs.** (MP(CPS) 104 (1988) 283–294, F MORICZ / BE RHOADES)

2881. $@/\mathcal{A}/\delta$ **'Riemann mapping Theorem' for Lorentz metrics** (PRSL A401 (1985) 117–130, RS KULKARNI)

2882. $\delta/\mathcal{A}/A$ **Stochastic complexity / statistical inference** (Bk: J Rissanen, World Sci., 1989) [See also: BLMS 23 (1991) 106–8]

2883. $\delta/@/A$ **Choquet Theory in Kantorovich Spaces** (Russ. Math. Surveys 30 (1975) 115–155, KUTATBLADZE, SS)

2884. $\delta/\mathcal{A}$ **Diagrammatic Techniques of Semigroup Th.** (PM HIGGINS, Techn. of 1/2-group Th., OUP, 1992)

2885. $\delta/\mathcal{A}$ **Additive #-Theory over** $F_q(T)$ (GW EFFINGER / DR HAYES, OUP 1991, pp 157) [∋ 'Vinogradov Theorem'...]

2886. $\delta/\mathcal{A}/@$ **Limit Theorems for Large Deviations** (L SAULIS / VA STATULE-VICIOUS, Kluwer, 1991, pp232)

2887. $\delta/\mathcal{A}/A$ **'Tracking Problems' (HS Scheme for Systems)** ('System Theory', A FEINTUCH / R SAEKS, AP (1982), pp310) [∋ machine tools: see BLMS 16 (1984) 440–2]

2888. $\delta/\mathcal{A}/A$ ***n*-var. pos.-real functions** (IRE Trans. Circ. Th. C77 (1960) 251–260, H OZAKI / T KASAMI & refs there)

2889. $\delta/@/A$ **Chem. reaction N-W** (IE^3 Trans. Circ. dist. CAS-21 (6) (1974) 709–721, G OSTER / A PERELSON)

2890. A/δ **Network Analogies** (FH BRANIN, Jr., in 'Large Scale Systems: 2', pp 357–362: GJ SAVAGE+, eds., Sandford Edu. P., 1978)

2891. $\delta/@/A$ **Frequency Spectra of TS** (Trans. Moscow MS (1981) (2) 163–200, ARHANGEL'SKII, Av.) [∋ complexity of closure operations]

2892. $\delta/\mathcal{A}/@$ **Topology / μ-Th. / Dyn. Syst. Th. / ...as specializations of operator-th.** (BLMS 14 (1982) 465–471: BE JOHNSON)

2893. $\delta/\mathcal{A}/@$ **Partial Integral-Eqs** (Bull. Res. Council Israel, 7F (1958), 181–6 S. KANTOROVITCH) [$\models$ reducibility to simul. Fredholm $\int$-eqs.]

2894. $\delta/\mathcal{A}/@$ **Géometrie Algébrique Réelle** (J BOCHNAK (M COSTE) Springer, 1987) [See BLMS 21 (5) 502–5)

2895. $\delta/\mathcal{A}/A$ **HS-Scheme for System Theory** (Feintuch, A / Saeks, R, AP, 1982 – cf. BLMS rev.)

2896. δ **Constructive Wiener-Hopf factorization** (PRSL A434, 419–433, DS JONES)

2897. δ **Forms of functional dependence** (AMM 74 (1967) 911-920*, WF NEMINS)

2898. $\delta/\mathcal{A}$ **Separation / Proximity Spaces** (AMM 74 (1964) 158–164: WJ PERVIN)

2899. δ **Algebraic Complexes / Chemical NW** (SIAM J. Appl. M. 15 (1967) 13–68, P SELLERS)

2900. δ **Measures of NW Passivity / Activity** (IE^3 Trans. Circ. Th. CT-17, 46–54, MR WALTERS (1970), AH SEMANiAN)

2901. δ/@ **Generalizations of 'Positive-Real FM'** (IE^3 CT-6 (1959) 374–383)

2902. δ **CV scheme for micropolar elasticity** (ZAMM 51 (1971) 183–188, ARIMAN / ZIKA)

2903. δ/@ **Operational Calculus for Finite Rings** (IE^3 CT-12 (1965) 558–570: J RICHALET)

2904. δ **Convolutions of highly singular GF** (PRSL A371 (1980) 471–508, DS JONES)

2905. δ **Constructive pf of connectedness of classical groups** (AMM 74 (1967) 964–6, Y-C WONG / Y-H Au-YEUNG)

2906. δ **Null-sets for Classes of Analytic Functions** (AMM 75 (1968) 462–470, L ZALCMAN)

2907. $\delta/\mathcal{A}$ **Statistical convergence in l.c. hausdorff spaces** (MPCPS 104 (1988), 141–5 IJ MADDOX)

2908. δ/@ **Topologies of MS of (formal) languages** (LNCS #53, 537–542, V VIANU)

2909. δ/@ **TS-Generalization of 'CAF'** (PLMS (3) 1 (1951) 152–162, P MAJSTENKO)

2910. $\delta/\mathcal{A}$ **Dist.** $[s, \{u_\lambda : u_\lambda \text{ uncontrollable}, \lambda \in \Lambda\}]$ (LNCIS #58, 303–314, R EISING) [Also: M WICKS et al., pp257–267 in RV PATEL++ eds., IEEE Repr. Vol (1994) : see pp215–267, for several 'closeness problems']

2911. δ **Nonstandard Analysis: Applications** (LMS Text #10, CUP, 1988, pp346, NJ CUTLAND, Ed.)

2912. δ/A **Converses in Information Theory** (MR 29 (1965) #1083)

2913. $\delta/\mathcal{A}$ **Measures on TS** (VARADARAHAN, VS, MR 26 #6842)

2914. $\delta/\mathcal{A}$ **k-Measures on Varieties** (FICHERA, G, ME 26 #6894)

2915. δ **1-D Phase Transition...** (JM Phys. 10 (1969) 1541–54, JL Strecker)

2916. δ/@ **Generalized invertible spaces** (AMM 73 (1966) 150–4, DX HONG)

2917. δ/@ **Universality of the S-function** (Mat. USSR Isv. 9 (1975) MR 57 #12419, 443–453, VORONIN, SM)

2918. $\delta/\mathcal{A}$ **Quasi-'Korovkin' criterion for GF** (See MR 36 (1968) #1975)

2919. δ **MS of ellipses / representation of noneuclidean geometries** (AMM 74 (1967) 673–7, LC NOVOA)

2920. δ **Discrete and Switching Functions** (DAVIO, M++, pp729, McGr-Hill, 1978, MR58 #15709)

2921. δ/𝒜 **'H/B of estimates in Th. of #'** (Spearman, B / Williams, KS, pp191, Carleton Math. LN #14 (1975) MR 53 #13084)

2922. δ **Systematic proofs of basic ineqs.** (AMM 76 (1969) 543–6, DE DAYKIN / CJ ELIEZER)

2923. **'Pointless Topology'** (BAMS (2) 8 (1983) 41–53, PT JOHNSTONE)

2924. **'Almost-FP' Theory** (Canad. JM 30 (4) 1978, 673–699, M HAZEWINKEL / M V.d. Vel, MR 58 #13004)

2925. **Converse Theorems for Lyapunov Stability** (Bol. Soc. Mat. Mex. 10/1959, 158–163, JL MASSERA *)

2926. **Infinitely Divisible Statistical Experiments** (LNST #27, Springer, 1985, A JANSSENS)

2927. δ **Categories of S-Rings: Extension of 'Group Theory'** (Symposia Math. Vol I, 5–13, 1969, O TAMASCHKE)

www.ingramcontent.com/pod-product-compliance
Lightning Source LLC
Chambersburg PA
CBHW081719170526
45167CB00009B/3633